The Stryker Brigade Combat Team

Rethinking Strategic Responsiveness and Assessing Deployment Options

Alan Vick • David Orletsky
Bruce Pirnie • Seth Jones

Prepared for the

United States Air Force

RAND

Project AIR FORCE

The research reported here was sponsored by the United States Air Force under Contract F49642-01-C-0003. Further information may be obtained from the Strategic Planning Division, Directorate of Plans, Hq USAF.

Library of Congress Cataloging-in-Publication Data

The Stryker Brigade combat team : rethinking strategic responsiveness and assessing deployment options / Alan Vick ... [et al.].
p. cm.
"MR-1606."
Includes bibliographical references.
ISBN 0-8330-3268-2
1. United States. Army—Reorganization. 2. Unified operations (Military science) 3. United States—Armed Forces—Foreign service. 4. Airlift, Military—United States. 5. Military sealift—United States. I. Vick, Alan.

UA25 .S857 2002
355.1'3'0973—dc21

2002031870

Top cover photo courtesy of U.S. Joint Forces Command at www.jfcom.mil. Photographer: PH2 Aaron Ansarov. Bottom cover photo courtesy of U.S. Air Force at www.af.mil/photos/transports.

Cover design by Stephen Bloodsworth

Published 2002 by RAND
1700 Main Street, P.O. Box 2138, Santa Monica, CA 90407-2138
1200 South Hayes Street, Arlington, VA 22202-5050
201 North Craig Street, Suite 202, Pittsburgh, PA 15213-1516
RAND URL: http://www.rand.org/
To order RAND documents or to obtain additional information, contact Distribution Services: Telephone: (310) 451-7002;
Fax: (310) 451-6915; Email: order@rand.org

PREFACE

Under the leadership of Chief of Staff General Eric Shinseki, the U.S. Army is taking steps to transform itself from the Legacy Force with its heavy forces, which are difficult to deploy strategically, and its light forces, which respond rapidly but lack staying power, into a more strategically responsive force. Its goal is to be able to respond quickly and decisively anywhere in the world. This transformation has begun with the creation of Stryker Brigade Combat Teams (SBCTs; formerly referred to as Interim Brigade Combat Teams), a brigade-sized force equipped with wheeled armored vehicles and other currently available weapons and vehicles. Intended for rapid deployment, the SBCT would move primarily by C-5 and C-17 aircraft. Over the next 20 years, General Shinseki envisions the entire Army combat force structure being replaced by the Objective Force, a brigade combat team capable of being deployed anywhere in the world in 96 hours after liftoff, with a division on the ground in 120 hours and five divisions in 30 days. The Objective Force will combine the deployability of light forces with the lethality, tactical mobility, and survivability of heavy forces. Both the SBCT and the Objective Force are expected to deploy a brigade anywhere in the world in 96 hours.

RAND's Project AIR FORCE conducted a study during fiscal year 2001 (FY01) of the implications of Army transformation efforts for U.S. Air Force (USAF) operations and force structure. Since the 96-hour deployment goal is central to the SBCT concept, the study assessed the requirements for rapid response and options for rapidly moving a brigade-sized force. This report seeks to answer the following questions: Can the Air Force meet the Army's 96-hour deployment goal? What combination of deployment and basing options would maximize the strategic responsiveness of new Army forces? How much unambiguous warning does the United States usually have before it initiates military operations? How much of this time will civilian decisionmakers typically consume in their deliberations before ordering deployment of military forces? Are large U.S. forces likely to deploy globally or just to certain regions? At what depths from the littoral might U.S. forces have to operate?

This study was conducted as part of the Strategy and Doctrine Program in RAND's Project AIR FORCE. It was sponsored by the Director of Operational Planning, Headquarters, USAF. It should be of interest to airmen and soldiers serving in plans, operational, analytic, and R&D organizations, as well as to the broader defense community. The information derived from it is current as of November 2002.

PROJECT AIR FORCE

Project AIR FORCE, a division of RAND, is the Air Force's federally funded research and development center (FFRDC) for studies and analysis. It provides the USAF with independent analysis of policy alternatives affecting the deployment, employment, combat readiness, and support of current and future air and space forces. Research is performed in four programs: Strategy and Doctrine; Aerospace Force Development; Manpower, Personnel, and Training; and Resource Management.

CONTENTS

FIGURES

TABLES

SUMMARY

BACKGROUND

Historically, to deter and defeat major threats in Europe and Asia, the United States has relied on forward-deployed Army and Air Force forces, Navy and Marine Corps forces afloat, long-range aircraft in the continental United States (CONUS), prepositioned unit sets in key regions, and reinforcing units from CONUS. For short-warning crises in other regions, Marine Expeditionary Units, the 82nd Airborne Division, Special Operations Forces, and Air Force/Navy air would be combined as appropriate to provide a limited capability that has usually been sufficient for noncombat evacuations and other lesser contingencies. The United States has not had the ability to deploy large joint forces globally from North America in a matter of days or weeks: The transportation challenge has been simply too great.

Army transformation efforts seek to establish that ability, turning the Army from the Legacy Force made up of well-equipped heavy warfighting forces, which are difficult to deploy strategically, and rapidly responding light forces, which lack staying power against heavy mechanized forces, to, first, an Interim Force of Stryker Brigade Combat Teams[1]—brigade-sized forces equipped with a

---

[1]The Interim Brigade Combat Team was renamed the Stryker Brigade Combat Team in August 2002. The brigade is built around the Stryker, a new infantry carrier vehicle named in honor of two Medal of Honor recipients: Private First Class Stuart S. Stryker and Specialist Robert F. Stryker, who served in World War II and Vietnam, respectively.

family of current-generation medium-weight wheeled armored vehicles—then, 15 to 20 years out, the Objective Force, equipped with the Future Combat System (FCS), a medium-weight tank that the Army hopes can be as survivable and lethal as the 70-ton M-1 tank.

To better understand the requirement for strategic responsiveness, as well as what is achievable, this study sought to answer the following questions:

- Can the Air Force meet the Army's 96-hour deployment goal?
- What combination of deployment and basing options would maximize the strategic responsiveness of new Army forces?
- How much unambiguous warning does the United States usually have before it initiates military operations?
- How much of this time will civilian decisionmakers typically consume in their deliberations before ordering deployment of military forces?
- Are large U.S. forces likely to deploy globally or just to certain regions?
- At what depths from the littoral might U.S. forces have to operate?

To assess deployment and basing options, the study team developed a simple spreadsheet that calculated transit times, loading and unloading times, and airfield throughput. It used military planning factors to determine aircraft maximum loads and ranges, and a variety of historical materials and interviews, as detailed in the Bibliography, to conduct the broader analysis of strategic responsiveness.

KEY FINDINGS

The main conclusion of this report is that a force with more than 1,000 vehicles cannot be deployed by air from CONUS to the far reaches of the globe in four days. However, with some mobility enhancements, it is possible to achieve deployment timelines on the order of one to two weeks, which is quite rapid for a motorized force.

Specifically, the combination of CONUS bases (particularly Fort Polk in Louisiana), an SBCT forward-based in Germany, and regional preposition sites in Guam and Diego Garcia offers the ability to deploy the SBCT by air or sea to key regions in 5 to 14 days. Figure S.1 illustrates specific times for scenarios in South America, East Asia, Africa, Europe, and Southwest Asia.

Findings for the other questions are as follows:

- Large U.S. joint operations have historically been concentrated in just a few regions: Europe, Latin America, the Persian Gulf, and Asia.
- The global war on terrorism is a wild card. It could lead to operations in locales more remote than the historical regions, but the forces involved for most such operations are likely to be special forces or other small, light forces, which are easy to deploy.
- Past security challenges (such as those that led to Operation Just Cause in Panama) have usually developed over a time frame of

RANDMR1606-S.1

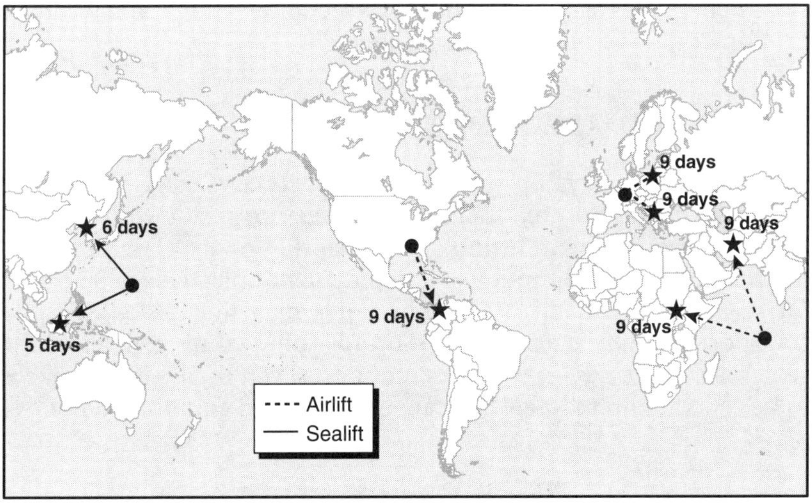

Figure S.1—Regional Basing, Showing Rapid Deployment Times to Key Areas

months or years, allowing for prepositioning and other regional defensive measures that reduce the need for rapid deployment from CONUS.

- Prepositioning of equipment or overseas basing of forces is the single most effective way to increase the responsiveness of U.S. Army forces for operations in key regions. From these preposition sites, the choice of airlift or sealift depends on the scenario.
- In general, deep interior deployments favor airlift; littoral scenarios favor sealift. Particular cases may present neither an airfield nor a port in proximity to the area of operations, requiring long road marches either way. To determine the preferred deployment mode, a detailed analysis of road networks and other local considerations would be necessary.
- For littoral deployments from preposition sites to ports in much of the Third World, fast, shallow-draft ships such as the catamaran ferries currently being tested by the Department of Defense (DoD) appear to offer the fastest and most robust option, although their shorter range may require more preposition sites than do larger ships. The ability of shallow-draft ships to use smaller ports avoids the time delays, complications, and weather constraints associated with lighterage.[2]

RECOMMENDATIONS FOR THE USAF

Army efforts to develop medium-weight forces[3] offer Joint Task Force or theater commanders capabilities not resident in current light or heavy forces and should be supported by the U.S. Air Force. Although the more-ambitious air deployment objectives of 96 hours for SBCTs and Objective Force brigades may not be feasible, air transport remains the fastest option for some contingencies. Fast sealift is promising for littoral operations. Even then, airlift is likely to be called upon to move critical personnel and equipment. For ex-

---

[2]*Lighterage* refers to the anchoring and unloading of large ships offshore, using landing craft (lighters) to carry the loads into the port.

[3]*Medium-weight forces* seek to combine the mobility and firepower of heavy forces with the deployability of light forces. The Stryker brigade truly is medium weight. It is half the weight of a heavy brigade and twice the weight of a light brigade.

ample, air might move Army port operations and security personnel and equipment to prepare a port for the arrival of the SBCT. Special forces might move by air to conduct supporting reconnaissance, direct action, or other special missions. In forced-entry scenarios, airborne forces might seize a port for the SBCT. Finally, airlift is likely to play a critical role in high-priority resupply and support operations.

More broadly, we note that the Air Force has a stake in Army transformation efforts. The Army envisions future forces operating in ways that are likely to require closer air-ground cooperation on intelligence, surveillance, and reconnaissance (ISR); lift; and precision fires. We recommend that Air Force and Army leaders initiate a dialogue on these issues of mutual concern. The Army would greatly benefit from the USAF's expertise on air deployment, ISR, survivability of transport aircraft, and air-to-ground fires. The USAF is beginning to develop new concepts for air-to-ground operations and would benefit greatly from Army expertise on land operations and from the substantial effort the Army has already invested in developing new concepts for the future battlefield.

ACKNOWLEDGMENTS

Lt Gen Howie Chandler, Commander of the 11th Air Force, was the original sponsor of the study. Emphasizing the importance of the mobility analysis, Gen Chandler encouraged us to make that analysis the focus of the study. Maj Gen Jeffrey Kohler took over sponsorship when he succeeded Gen Chandler as Director of Operational Plans and Joint Matters, Headquarters, U.S. Air Force (HQ USAF). Gen Kohler provided valuable guidance and feedback throughout the study. We greatly appreciate the support, enthusiasm, and encouragement that both our sponsors offered the study. Lt Col Mark Hicks, HQ USAF/XOXS, was the project monitor. After Col Hicks' reassignment to Air Combat Command, Maj Michael Pietrucha took over as project monitor. Both superbly supported the project on substantive and administrative matters. We thank Maj Pietrucha and Lt Col Gary Crowder, HQ Air Combat Command, for their trenchant comments on the draft report.

Dr. Arnold Kanter (Under Secretary of State, 1991–1993) and Morton Halperin (Director, Policy Planning Staff, Department of State, 1998–2001) graciously agreed to be interviewed and shared their insights on crisis decisionmaking and the decision to use military force.

Maj Gen Ron Bath, Director of Strategic Planning, HQ USAF, offered many helpful comments on our briefing and kindly shared his own briefing on transformation issues.

Col Ron Richardson, Director of Homeland Security, HQ USAF, shared his experience from his tour commanding the 437th Air Expeditionary Group during Operation Allied Force. Col Richardson also provided valuable data from that operation.

BG Charles Fletcher, then Director, Force Projection and Distribution, Office of the Deputy Chief of Staff for Logistics, Headquarters, U.S. Army (HQ USA), invited us to see his analysis of SBCT deployment options. We thank him for generously spending several hours with us discussing the Army analysis.

We thank Jonathan Kaskin, Director of Strategic Mobility and Combat Logistics Division (N42), Headquarters, U.S. Navy (HQ USN), and CAPT Mike Brannon, USN, for discussing fast sealift options and challenges.

Col Rob Owen, HQ Air Mobility Command (AMC), hosted a study visit to Scott AFB, Illinois, and offered a stimulating historical perspective on the challenges of tactical airlift. At AMC, we also thank Robert Dangredo, Dave Merrill, and Col Mike Fricano. At U.S. Transportation Command, we thank COL Arthur Breithaupt, USA.

Maj Scott Smith, HQ USAF, shared his master's thesis on deploying Army forces by air and offered valuable comments on our briefing.

Philippe Loustaunau and Mark Lewis at the Institute for Defense Analyses kindly shared their research on new concepts for rapid deployment and commented on our briefing.

Richard Lowrie, INCAT USA, provided information on INCAT's fleet of fast catamaran ferries.

At RAND, we thank John Halliday and Jed Peters for their incisive and constructive reviews of our draft report. We also thank project colleagues Jeff Marquis and Sean Edwards for their contributions to the study. We also worked closely with colleagues in the Arroyo Center who provided insights and expertise on the Army Stryker Brigade Combat Team. We thank Tom McNaugher for his support and encouragement and John Gordon, Dan Norton, Eric Peltz, David Kassing, and Walt Perry for sharing their analyses and perspectives. MAJ Patrick Kirk, RAND Army Fellow, provided us the latest information on the SBCT. We thank Rob Button, Frank LaCroix, Carl Stephens, Mahyar Amouzegar, and Geoffrey Sommer for sharing their expertise and references on naval operations, particularly fast sealift. Ted Harshberger, Bob Roll, Bob Tripp, Don Palmer, Glenn Kent, Dave Ochmanek, and Bruce Nardulli all provided helpful comments on briefings of this work.

Finally, we thank Leslie Thornton and Joanna Alberdeston for their skilled preparation of the manuscript; Marian Branch for her thorough editing; and Stephen Bloodsworth for the stunning cover design.

ACRONYMS

AB	Airbase
ACL	Allowable cabin load
AEF	Air Expeditionary Force
AEW	Air Expeditionary Wing
AFB	Air Force Base
AFSOC	Air Force Special Operations Command
AMC	Air Mobility Command
APOD	Aerial port of debarkation
APOE	Aerial port of embarkation
ASG	*Abu Sayyaf* Group
AUC	*Autodefensas Unidas de Colombia*
AWR	Army War Reserve
CENTCOM	Central Command
CIS	Commonwealth of Independent States
CONUS	Continental United States
CRAF	Civil Reserve Air Fleet
DARPA	Defense Advanced Research Projects Agency

DoD	Department of Defense
DPRK	Democratic People's Republic of Korea (North Korea)
EAPC	Euro-Atlantic Partnership Council
ELN	*Ejercito de Liberación*
FARC	*Fuerzas Armadas Revolucionarias de Colombia*
FCS	Future Combat System
FM	Field Manual
FOD	Foreign object damage
FSS	Fast Sealift Ships
HMMWV	High-Mobility, Multipurpose, Wheeled Vehicle
IAV	Interim Armored Vehicle
ISR	Intelligence, surveillance, and reconnaissance
JCS	Joint Chiefs of Staff
JLOTS	Joint Logistics Over-the-Shore
JV 2020	*Joint Vision 2020*
LAV	Light Armored Vehicle
LCM	Landing Craft Mechanized
LCU	Landing Craft Utility
LMSR	Large, Medium Speed RoRo
LSV	Landing Support Vessel
MAP	Membership Action Plan (NATO)
MILF	Moro Islamic Liberation Front
MNLF	Moro National Liberation Front

MOG	Maximum on ground
MPS	Maritime Prepositioning Ships
MSC	Military Sealift Command
NAC	North Atlantic Council
NALMEB	Norway Air-landed Marine Expeditionary Brigade
NCA	National Command Authority
NGO	Non-governmental organization
NIMA	National Imagery and Mapping Agency
OPFOR	Opposing force
OSCE	Organization for Security and Cooperation in Europe
PAA	Primary Aircraft Authorized
PAI	Primary Authorized Inventory
PAX	Passengers
PfP	Partnership for Peace
PMAI	Primary Mission Aircraft Inventory
RoRo	Roll-on/roll-off (ship)
ROS-4	Reduced operational status of 4 days
RRDF	Roll-on/roll-off discharge facility
SBCT	Stryker Brigade Combat Team (formerly Interim Brigade Combat Team)
SLWT	Side-loadable warping tug
SPOD	Sea port of debarkation
SPOE	Sea port of embarkation
SSC	Small-scale contingency

TAI	Total aircraft inventory
TO&E	Table of Organization and Equipment
U.K.	United Kingdom
UN	United Nations
UNOSOMI	First UN Operation in Somalia
UNPROFOR	United Nations Protection Force
UNSC	United Nations Security Council
UNSCOM	United Nations Special Commission
UNTAET	United Nations Transitional Authority in East Timor
USA	United States Army
USAF	United States Air Force
USMC	United States Marine Corps
USN	United States Navy
UTE	Utilization rate
WMD	Weapons of mass destruction

Chapter One

INTRODUCTION

Historically, to deter and defeat major threats in Europe and Asia, the United States has relied on forward-deployed Army and U.S. Air Force (USAF) forces, Navy and United States Marine Corps (USMC) forces afloat, long-range aircraft in the continental United States (CONUS), prepositioned unit sets in key regions, and reinforcing units from CONUS. For short-warning crises in other regions, Marine Expeditionary Units, the 82nd Airborne Division, Special Operations Forces, and USAF/Navy air would be combined as appropriate to provide a limited capability that was usually sufficient for noncombat evacuations and other lesser contingencies. The United States has not had the ability to deploy large joint forces[1] globally from North America in a matter of days or weeks: The transportation challenge has been simply too great.

The U.S. Army is undergoing a transformation that, theoretically, will establish that ability. As described in detail in the next section, the Army's Legacy Force of well-equipped, heavy warfighting forces, which are difficult to deploy strategically, and of rapidly responding light forces, which lack staying power against heavy mechanized forces, are being supplemented by an Interim Force of medium-sized—brigade-sized—teams equipped with medium-weight armored vehicles that the Army wants to deploy anywhere in the world in 96 hours after liftoff. This concept will evolve into the Objective

[1]We define a *large joint force* as deploying at least one ground force brigade equivalent and one air wing equivalent. Such a force might include an army brigade (e.g., the Stryker Brigade Combat Team [SBCT]), a Marine Expeditionary Unit, a USAF air expeditionary wing, a Carrier Battle Group, and an Amphibious Ready Group.

Force, which has the objective of deploying a medium-sized force (brigade) within 96 hours, a division a day later, and five divisions by month's end.

BACKGROUND

The seeds for current Army transformation efforts were sown during two recent conflicts in which there was a need for heavy forces to deploy fairly quickly. During Operation Desert Shield, the United States sought to rapidly move sufficient force to the Persian Gulf to defend Saudi Arabia from Iraqi armored forces poised on the Saudi-Kuwaiti border. No Army force at the time was both sufficiently light to move rapidly by air and in possession of the lethality, survivability, and mobility to stop Iraqi armor. As a stopgap measure, the 82nd Airborne Division was deployed. A relatively light unit[2] with limited mobility and limited anti-armor capability, this division was the only U.S. ground combat force standing between Iraqi heavy divisions occupying Kuwait and the oil fields and cities of Saudi Arabia.

After the Persian Gulf War, many in the Army expressed considerable disquiet over the lack of a rapidly deployable force that could stop enemy armored forces. Postwar analyses conducted at RAND and elsewhere suggested that the 82nd Airborne would not have been able to stop Iraqi heavy forces if it had continued its offensive into Saudi Arabia.[3]

The second conflict, Operation Allied Force, the 1999 NATO air operation to compel Serbia to withdraw its forces from Kosovo, also highlighted the need for rapidly deployable, lethal, and mobile Army forces. U.S. planners confronted a wide gap between an air-only effort, which could start almost immediately, and an air-ground effort, which would take months to prepare, especially considering the highly constricted lines of communication. It would have been too risky to employ airborne and airmobile forces against the Serbs, who

---

[2]It was somewhat heavier than a "light" brigade because of vehicles, and much lighter than an armored division.

[3]See, for example, John Matsumura et al., *Lightning Over Water: Sharpening America's Light Forces for Rapid-Reaction Missions*, Santa Monica, Calif.: RAND, MR-1196-A/OSD, 2000.

had main battle tanks and other armored vehicles. But heavy forces would have had to disembark either at Durrës on the Adriatic Sea and march through Albania or at Thessaloniki on the Aegean Sea and march through Greece and Macedonia. According to informal plans, United Kingdom (U.K.) forces were to advance through Macedonia and U.S. forces, through Albania. By the time Serbia capitulated to the NATO demands, U.S. engineers had reconnoitered routes through Albania, finding steep gradients, narrow shoulders, antiquated bridges, and narrow defiles through the mountains between Albania and Kosovo—daunting obstacles for the movement of heavy forces.

At the direction of General Wesley K. Clark, Supreme Allied Commander, Europe, and Commander in Chief, U.S. European Command, the United States airlifted Task Force Hawk to Albania during Operation Allied Force. To protect its AH-64 helicopters from attack by Serb forces based a short distance away in Montenegro, Task Force Hawk included heavy forces equipped with Abrams main battle tanks and Bradley infantry fighting vehicles. C-17 and C-130 aircraft delivered these Task Force Hawk forces to Rinas Airport, a poorly developed airfield already crowded with humanitarian flights.[4] This airlift was an impressive debut for the C-17, but Task Force Hawk was not employed,[5] except in conducting reconnaissance with unmanned aerial vehicles and finding Serb artillery posi-

---

[4]Moving V Corps involved 442 C-17 sorties and 269 C-130 sorties. Moving elements of the 82nd Airborne Division required 26 C-17 sorties. See Headquarters, U.S. Air Force Europe/AMOCC, "Allied Force: Intra-Theater Air Mobility," briefing, prepared by MAJ Martin Beard, USA, July 1999.

[5]The debate over whether to employ Task Force Hawk occurred in the context of a wider debate (both political and military) over whether to invade Serbia if Belgrade continued to refuse NATO's demands. With regard to Task Force Hawk, the Army raised both policy and operational concerns; however, the first question was how many losses attack helicopters might suffer. The Task Force Hawk commander, Lt Gen Jay Hendrix, estimated losses of 0–5 percent, but stressed that accurate estimation was impossible. Gen Wesley K. Clark, the Supreme Allied Commander, Europe, was dismayed that the Army did not support employment of its helicopters. Another problem was suppressive fire against enemy air defense, which raised the issue of collateral damage. As to an invasion, Gen Clark advocated preparing for simultaneous advances through Albania and Macedonia to clear Kosovo of Serb forces, but encountered so much skepticism from the Joint Chiefs of Staff that he avoided asking for a decision. The Department of Defense deliberately kept him away from higher-level discussions of this issue. See Wesley K. Clark, *Waging Modern War: Bosnia, Kosovo, and the Future of Combat,* New York: Public Affairs, 2001, pp. 162–344, 430–437.

tions by radar. Serb leaders may have perceived the task force as a precursor to invasion and therefore concluded (correctly) that NATO would invade if necessary to expel Serb forces from Kosovo. Serb leaders also knew that the task force was too small to threaten them directly.[6]

The Army has since begun a transformation that seeks to offer U.S. leaders better options in future Kosovos.

ARMY TRANSFORMATION INITIATIVES

The Army transformation will affect every aspect of its doctrine, training, organization, and equipment. Chief of Staff General Eric K. Shinseki explains the rationale in an October 2000 article in *Army*:[7]

> Our legacy Army's warfighting prowess today is assembled around two force characteristics—*heavy* and *light*: magnificent *heavy* forces that are well equipped for war but difficult to deploy strategically, and magnificent *light* forces that can respond rapidly and are well suited for stability and support operations but lack staying power against heavy mechanized forces. . . . With each passing year, our condition as a force becomes a greater liability.

The Army is currently fielding the Stryker brigades, which are equipped with medium-weight armored vehicles, and which, according to General Shinseki,[8] "will meet an operational shortfall that currently exists between the capabilities of our early arriving light forces and our later arriving heavy forces." The Stryker brigade is envisioned as the precursor of an Objective Force, also medium weight,[9] which should start to become available before the end of this decade. If the Objective Force were successful, the entire Army,

---

[6]See Bruce Nardulli et al., *Disjointed War: Military Operations in Kosovo, 1999*, Santa Monica, Calif.: RAND, MR-1406-A, 2002, especially pp. 62–79.

[7]General Eric K. Shinseki, "The Army Transformation: A Historic Opportunity," *Army*, October 2000, p. 23.

[8]General Eric K. Shinseki, "The Army Vision: A Status Report," *Army*, October 2001, p. 33.

[9]*Medium-weight forces* seek to combine the mobility and firepower of heavy forces with the deployability of light forces. The Stryker brigade truly is medium weight. It is half the weight of a heavy brigade and twice the weight of a light brigade.

less only highly specialized units, would eventually be transformed to this same design.

As background for the analysis that follows, we look at the different transformation initiatives, beginning with emerging joint-force doctrine.

Emerging Doctrine

Joint Vision 2020 (JV 2020) sets the context for new Army doctrine and introduces the term *transformation* on page 1:

> If our Armed Forces are to be faster, more lethal and more precise in 2020 than they are today, we must continue to invest in and develop new military capabilities. This vision describes the ongoing transformation to those new capabilities.[10]

JV 2020 defines two concepts central to the Army's emerging doctrine: dominant maneuver and precision engagement.

Dominant maneuver means gaining positional advantage through decisive speed and overwhelming operational tempo. Moreover, dominant maneuver is envisioned on a global scale: "Overseas or US-based units will mass forces and effects directly to the operational theater."[11] For the Army, dominant maneuver implies much more rapid arrival in theater than had been achieved previously. It also implies that ground combat units must arrive ready to fight without the usual reception, staging, and preparation.

Precision engagement is the ability to locate and track targets, to engage targets with appropriate systems, and to achieve the desired effects. It implies a system of systems that links sensors and delivery systems. For the Army, precision engagement implies a networked family of combat vehicles capable of combined arms at lower levels of command than in the past, plus better connectivity to systems operated by sister services.

---

[10]Chairman, Joint Chiefs of Staff (CJCS), *Joint Vision 2020*, Washington, D.C.: J-5, Strategic Plans and Policy, June 2000, p. 1.

[11]CJCS, 2000, p. 20.

The Army's primary manual at the operational level, Field Manual 3-0, *Operations*,[12] emphasizes timely deployment of land forces:

> Commanders view force projection as a race between friendly forces and the enemy or situation. The side that achieves a decisive operational capability first seizes the initiative.[13]

In its introduction to Part Two, it states five general rules: "Army forces win on the offensive; initiate combat on their terms . . . ; gain and maintain the initiative; build momentum quickly; and win decisively."[14] To develop combat power, land forces employ maneuver and firepower. *Maneuver* is defined as employment of forces through movement combined with fire to achieve a position of advantage. *Firepower* is defined as the destructive force essential to overcoming an enemy's ability and will to fight. Maneuver and firepower complement and magnify each other. Neither is decisive in isolation; combined, they ensure an enemy's defeat.

The Stryker Brigades

The Army Vision document defines *transformation* as a threefold effort encompassing the Legacy Force, the Stryker brigades, and the Objective Force. The Legacy Force is today's Army, characterized by two extremes: one light and the other heavy. At the light extreme, forces include light infantry divisions (10th Mountain Division, 25th Infantry Division), an airborne division (82nd Airborne Division), and an airmobile division (101st Airborne Division), which are essentially foot infantry once they enter combat. The heavy extreme includes such forces as those forward-deployed in Europe (1st Infantry Division [Mechanized], 1st Armored Division), which means that they are equipped with the Abrams tank and the Bradley infantry fighting vehicle.

---

[12]Headquarters, Department of the Army, Washington, D.C., June 14, 2001. Traditionally known as FM 100-5, this manual was designated FM 3-0 to agree with the joint numbering system.

[13]Headquarters, Department of the Army, 2001d, paragraph 3-43.

[14]Headquarters, Department of the Army, 2001d, Introduction to Part Two: Foundations of Full Spectrum Operations.

Army leadership considers there to be a dangerous gap between these two force extremes:

> More than ten years ago, during the buildup of Operation Desert Shield, the Army identified an operational shortfall—a gap between the capabilities of our heavy and light forces. Our heavy forces are the most formidable in the world. There are none better suited for high-intensity operations, but they are severely challenged to deploy to all the places where they might be needed. Conversely, our magnificent light forces are agile and deployable. They are particularly well suited for low-intensity operations, but lack sufficient lethality and survivability. There is, at present, no rapidly deployable force with the staying power to provide our leadership with a complete range of strategic options.[15]

The Stryker brigades fill this gap in the near term while providing an organization for developing concepts for the Objective Force. There will be six Stryker Brigade Combat Teams[16] (SBCTs). The first SBCT, the 3rd Brigade of the 2nd Infantry Division at Fort Lewis, Washington, will be operational in early 2003. The second SBCT, the 1st Brigade of the 25th Infantry Division (Light), also at Fort Lewis, will be operational in 2004.[17] These units are equipped with a family of Interim Armored Vehicles (IAVs) built on the commercially available Light Armored Vehicle (LAV; manufactured by General Motors, Canada, and General Dynamics). IAV variants are optimized for the following purposes: infantry carrier, reconnaissance, anti-tank, mortar, command, fire support, engineer, nuclear-biological-chemical protection, and medical support. A Mobile Gun System, currently under development, will also be part of the unit equipment.

---

[15]Thomas E. White, Secretary of the Army, and General Eric K. Shinseki, *Joint Statement Before the Committee on Appropriations Subcommittee on Defense on the Fiscal Year 2002 Army Budget*, Washington, D.C.: U.S. Senate, June 13, 2001, p. 3.

[16]The Interim Brigade Combat Team (IBCT) was renamed the Stryker Brigade Combat Team in August 2002. The brigade is built around the Stryker, a new infantry carrier vehicle named in honor of two Medal of Honor recipients: Private First Class Stuart S. Stryker and Specialist Robert F. Stryker, who served in World War II and Vietnam, respectively.

[17]The remaining four brigades are the 172nd Infantry Brigade (Separate) in Fort Wainwright, Alaska; the 2nd Armored Cavalry Regiment (Light) in Fort Polk, Louisiana; the 2nd Brigade of the 25th Infantry Division (Light) in Schofield Barracks, Hawaii; and the 56th Brigade (Mechanized) of the 28th Infantry Division (Mechanized). All except the 56th Brigade, which is in the Pennsylvania National Guard, are active Army units.

The SBCT is designed for rapid deployment, its goal "to place a credible combat force on the ground anywhere in the world in 96 hours from liftoff."[18] Over long distances, the SBCT would move primarily by C-5 and C-17 aircraft. It might also be forward-deployed or moved by combinations of airlift and sealift. Within a theater of operations, it could deploy by C-130 aircraft, by its own vehicles, by rail, or by other means.[19] All IAVs organic to the SBCT must be transportable by C-130 and must be able to enter and exit the aircraft capable of conducting immediate combat operations, although not necessarily carrying full basic loads. Their combat-capable weights must not exceed 19 short tons.

The SBCT is configured to arrive early in a crisis but is not an assault force. It would normally deploy to an airfield, airstrip, or seaport under friendly control. Thus, in some scenarios, Marines, the 82nd Airborne, or Army Rangers would have to first seize an airfield or port for the SBCT to use. To facilitate rapid deployment, the SBCT would ordinarily deploy with a basic load sufficient for a few days in combat, after which it would have to be supplied from higher echelons. In combat, the SBCT would assault as dismounted infantry, although it might take some immediate actions without dismounting. The IAV must provide all-around protection from small arms and accept add-on armor to defeat heavy machine-gun fire and handheld rocket-propelled grenades.

The Objective Force

In its concept for the Objective Force, the Army stresses rapid deployment using advanced airlift. It sets the following goals: "The Army goal is to deploy a brigade combat team anywhere in the world in 96 hours after liftoff, a division on the ground in 120 hours, and five divisions in 30 days."[20] The Army expects the Objective Force to

---

[18]Headquarters, Department of the Army, *Operational Requirements Document for a Family of Interim Armored Vehicles (IAV), Prepared for Milestone I Decision*, Washington, D.C., April 6, 2000c, p. 2.

[19]Headquarters, Department of the Army, 2000c, p. 23.

[20]Headquarters, Department of the Army, *Concept for the Objective Force*, Washington, D.C.: White Paper with foreword by General Eric K. Shinseki, Chief of Staff, November 2001c, p. 9.

provide attractive options to joint force commanders that will justify allocation of scarce lift assets to deployment of Army forces. It recognizes that sustaining Army forces in combat will be the most challenging requirement for lift. In designing the Objective Force, it intends to reduce and pace demands for consumables, especially ammunition and fuel.

The Army initially conceived of the Objective Force as the product of an extensive effort in research and development for which the science and technology community would develop a research plan by 2003. The Army would make technology investments to realize that plan, and, after 8–10 years of development, the design of the Objective Force would be achieved.[21] At the heart of the Objective Force will be the Future Combat System (FCS), currently being explored in a joint program by the Army and the Defense Advanced Research Projects Agency (DARPA). The Army recently decided to accelerate FCS development so that the first unit would be equipped in 2008 and initial operational capability would be attained by 2010, which means that technologies will need to be mature in 2003 if they are to affect the initial configuration of the FCS. The Army anticipates providing technology-insertion points for subsequent upgrades. Currently, the Army has set just one firm specification for the FCS: It must be transportable by C-130 aircraft.

Although the FCS is constrained to less than 20 tons, the Army expects the Objective Force to close with and destroy enemy forces in mounted combat as heavy forces do today. To accomplish this task at an acceptable risk to friendly forces will demand innovative tactics and excellent situational awareness. The Objective Force is conceived as follows:

- Fully networked to enable highly flexible combined-arms operations at battalion and company levels
- Equipped with several classes of unmanned aerial vehicles for reconnaissance

---

[21]Shinseki, 2000, p. 28.

- Probably outfitted with robotic land vehicles for use in close combat, at least for reconnaissance and perhaps also for engagement of targets.

Reconnaissance receives heavy emphasis. Whereas, in the past, Army forces have usually sought contact with enemy forces, compelling them to reveal their positions and capabilities through combat, in the future, the Army hopes to attain a comprehensive picture of enemy forces *before* going into direct combat. Such a picture might allow the Army to defeat enemy forces with standoff and indirect-fire weapons rather than engage in more dangerous close-in fights. It is, as yet, unclear whether the FCS would be mated with some future aircraft—for example, a large tilt-rotorcraft or tilt-wing aircraft—that has a greater ability to land on unimproved strips or even open ground. While this option remains attractive to the Army, it is not central to the Objective Force design.

PURPOSE AND ORGANIZATION

This study was originally intended to help the USAF better understand the broad implications of Army transformation efforts for air power. In the course of the research, it became evident that *strategic responsiveness*[22] was a critical issue and one that deserved more analytic attention. Therefore, this report focuses on the challenge of moving a motorized Army brigade, such as the SBCT, and

- assesses the contributions of airlift, sealift, and prepositioning to the solution. Chapter Two uses DoD planning factors for this assessment.
- puts responsiveness in a broader context by considering the timelines associated with the political-military decision to employ force. Using histories of past operations, Chapter Three presents an analysis of political-military decisions to initiate military operations that employ joint forces. It examines how much warning time is likely and how leaders make the decision to deploy forces.

---

[22]We define *strategic responsiveness* as the ability of a force to deploy decisive combat power on a timetable that supports the objectives of U.S. leaders.

- identifies the regions where U.S. military forces are most likely to be deployed. Chapter Four analyzes past operations, accessibility, and security issues that might trigger future operations, to identify these regions and to determine whether brigade or larger joint forces really need to deploy anywhere in the world or whether a regional focus is more appropriate.

Chapter Five presents our conclusions and recommendations. There are also two appendices. Appendix A describes historical depths of U.S. sealift operations from the littoral. Appendix B tabulates the components of airlift and sealift deployment times for scenarios in Chapter Two.

Chapter Two

DEPLOYING THE SBCTs

Many factors affect the time required to deploy units, including the location of deploying units; nearness to airfields or ports; the departure point; overflight rights; location of en route airbases (if such airbases are required); mobility assets allocated to deployment (i.e., number and types of airlifters, tankers, and cargo ships); composition of the SBCT; suitability of host-nation infrastructure (airports and seaports); quality of road network from port of debarkation to deployment location; and weather.[1] The SBCT deployment time will depend heavily on the specifics of the situation and will require detailed operational analysis to arrive at accurate answers. The objective of this analysis is to obtain an understanding of the big issues associated with the deployment of the SBCT throughout the world. Our analysis used mobility planning factors (a combination of airlift and sealift planning factors)[2] to analyze a range of scenarios to better understand the important aspects of global mobility as it pertains to SBCT deployment. Planning factors allow many cases and scenarios

---

[1]The recent movement of units to Kosovo provides an example of the importance of scenario "ground rules" on outcome. Considering the airbase at Tiranë/Rinas in Albania, one could expect throughput on the order of 4 C-17s maximum on ground (MOG). However, the requirement for humanitarian relief supplies meant that only half of this MOG was allocated to the deployment of military units, with the other half going to non-governmental organizations (NGOs) providing the relief supplies. In many scenarios, this split would double the time required to deploy a unit. (See the Airport Infrastructure subsection for a discussion of MOG.)

[2]*Airlift planning factors* include average loading and unloading times, aircraft payload in tons or number of personnel, average time on ground, aircraft unrefueled range for a particular payload, crew rest requirements, and number of hours the aircraft can be flown per day. *Sealift planning factors* include loading and unloading times, payload, and docking/port maneuvering time.

to be explored quickly, offering a fairly simple approach to obtaining gross estimates on the time required to deploy units. The methodology covers the major aspects of the deployment, including the number and types of aircraft allocated to unit deployment, deployment distance, the contingency USE rate[3] of aircraft, airfield-throughput constraints, aircraft payload, and aircraft speed.

The following analysis provides comparisons between airlift and sealift deployments. The *airlift deployment time* is "wheels-up" time, which *starts* when the first aircraft is airborne from the departure airport and *ends* when the last aircraft needed to deploy the unit touches down at the arrival airport. It is strictly for the deployment of the SBCT without augmentation or additional support and does not include time to position and prepare the mobility system.

A similar measure, *sealift deployment time,* is used for sealift. It is calculated from the moment the ship begins to load at the port of embarkation to when the ship is completely unloaded at the port of debarkation.[4]

The time prior to wheels-up for aircraft or prior to loading of the ships (which is likely to be 2 to 4 days, depending on the situation), is the period required to assemble the SBCT and move to the airport or seaport, recall the airlift crew from their daily operations, position the crews at en route bases for crew rest, or activate the sealift.[5]

---

[3]*Contingency USE rate* is the capability of a subset of Primary Mission Aircraft Inventory (PMAI) aircraft to generate flying hours. It is expressed in average flying hours per day (e.g., 11.7 hours for the C-17). UTE rate is a related and better-known metric, but not appropriate for this analysis. UTE rate calculations are based on Primary Authorized Inventory (PAI) aircraft and apply only to long-term, large-scale operations that require 100-percent Active and Reserve participation. As noted earlier, the SBCT is optimized for small-scale contingencies, which will generally not meet UTE-rate criteria. For more details, see AFPAM 10-1403 (HQ USAF, 1998).

[4]Loading/unloading time is included in the sealift deployment time, but not in the airlift deployment, because loading the first airlift sortie and unloading the last sortie will take a few hours. The sealift loading/unloading time, by contrast, will take days, as is shown later in this chapter.

[5]Preparing the airlift system for a major deployment is not a trivial task. First, airlift and aircrews must be recalled from their peacetime missions and positioned to the theater before the start of the deployment. En route bases must be identified. Political constraints must be met. Aircrews, maintenance personnel, and equipment must be prepositioned. If tankers are required, these functions must also be performed for the tanker aircraft prior to beginning the deployment.

This chapter assesses the ability of air power to achieve the Army's goal of global deployment for an SBCT in 96 hours, considers the relative advantages of airlift, sealift, and forward prepositioning, and identifies some of the more promising options for enhancing the strategic mobility of the SBCT and other motorized Army forces. The first section of the chapter uses USAF airlift planning factors to determine how quickly the SBCT can be deployed to various distances. The second section assesses SBCT deployment by sea and emphasizes particularly the problems associated with deploying to small ports in the developing world. The last section compares these deployment options for four different scenarios.

AIRLIFTING THE SBCT

The first step in SBCT deployment analysis is determining the overall size of the unit. Although uncertainty still exists regarding the number and types of vehicles that will make up the SBCT, enough information is available to estimate the airlift deployment requirements.

Earlier, unpublished RAND work conducted by Daniel M. Norton of the Arroyo Center[6] provided a basis for our estimates of the amount of equipment and personnel that must be deployed as part of the SBCT. Using the information provided by the Army on vehicles, equipment, and personnel, the Arroyo Center project estimated the weight of the equipment and number of personnel required for the deployment effort. Sustainment for these forces must also be deployed, and estimated. *Sustainment* covers all of the supplies required to conduct combat operations and provide food, water, and shelter for the forces. Usually, fuel and water are the two heaviest and most difficult sustainment items to deploy. Our calculations (and those of the RAND Arroyo Center project) assume that the host nation will provide fuel and water, thereby greatly reducing the lift burden for the initial deployment, as well as the lift required throughout the operation. However, this assumption may not be realistic for all scenarios.[7] The Army intends to deploy the SBCT with

---

[6]Daniel M. Norton, "IBCT Deployment: An Initial Assessment," Santa Monica, Calif.: unpublished RAND research.

[7]For example, the USMC Brigade Task Force that deployed to Camp Rhino, south of Kandahar, Afghanistan, had to bring all fuel and water with it.

ammunition and other stocks sufficient for 3 days of combat. We used 3 days in our analysis as the absolute minimum that would be required as part of the initial deployment. Presumably, additional stocks would be flown in immediately so that the SBCT could quickly build up more than a 3-day supply of ammunition and other essential stocks.[8]

A final item that must be considered is the deployment of the equipment and personnel required for the airlift operation itself—a "mobility package." Since the SBCT is optimized for small-scale contingencies (SSCs), the deployment airports likely to be used may be austere and neglected. Even if the airports were at the highest standard prior to the conflict, they may have suffered some combat damage. Damage to runways, lights, taxiways, and other critical infrastructure would need to be repaired before high-tempo, efficient, and safe air operations could begin. In any event, Air Force personnel and equipment are required to prepare the airport for operations and provide adequate force protection. Although some of these items may be required only during the initial phase of the operation (for example, repair of landing lights and runways), these items must be deployed initially as part of the operation.

Our analysis uses the same mobility package as that developed for an Air Expeditionary Wing (AEW) deployment.[9] Many of the same personnel and much of the equipment would be required at any airfield, since many of the functions required to prepare and sustain mobility operations are independent of the unit being deployed. Of course, some components of the mobility package will be different (for example, aircraft munition loaders will not be required for the SBCT), but an AEW mobility package provides a good initial estimate of elements required for SBCT deployment. Table 2.1 presents our esti-

---

[8]Assuming that the SBCT can deploy with only 3 days of sustainment is optimistic. Any disruption of the sustainment flow could result in shortages. Further, this estimate does not account for the sustainment that will be consumed throughout the deployment operation by components of the unit that are deployed first, which will result in less than 3 days of sustainment actually being available at the end of the deployment operation.

[9]HQ USAF, Office of the Deputy Chief of Staff for Air and Space Operations, "AEF Sustainment Estimates," unpublished briefing, 26 September 2000.

Table 2.1

SBCT Initial Deployment Requirement

Unit Element	Weight (short tons)[a]	Personnel
SBCT	12,840	3,494
3-day sustainment (no fuel or water)	2,500	
Mobility package	900	1,030

[a]These weights for the SBCT were derived from the IBCT Table of Organization and Equipment (TO&E) in 2001. Since then, the TO&E has gone through several changes. An April 2002 study by the U.S. Transportation Command showed the brigade growing to 14,663 short tons and 3,863 personnel. In summer 2002, Army Major Lawrence Raville (while on a tour as a Research Fellow at RAND) calculated the SBCT to weigh in at 13,554 short tons and 3,418 passengers. The final weight for the brigade remains uncertain. See U.S. Transportation Command, *Interim Brigade Combat Team Air Mobility Deployment Analysis*, Scott AFB, Ill., April 2002, p. 2.

mate of the personnel and equipment weight required for SBCT deployment.

Many factors determine the number of aircraft loads required to deploy a unit. First, deployment of a unit by air depends on the size as well as the weight of the unit. Some items are light but bulky and cause the aircraft to be filled before it has reached its maximum weight; other items cause the aircraft to reach its maximum desired weight before its cargo space is completely filled. How efficiently a particular unit loads aircraft is driven largely by whether the loading requirement is operational or administrative. Units that are *operationally loaded* will require more airlift, since they load in the order required for combat, not in the order that would make most efficient use of the aircraft. An operational load is required when the units may engage in action shortly after reaching the destination; an *administrative load* assumes that the unit will have the time required to get all of its equipment, personnel, and sustainment to the destination prior to engaging in any action. Our calculations are based on planning factors that reflect historical loadings, which are usually administrative and may overstate what could be carried in those situations for which units need to come off the aircraft in combat formation.

Another constraint is associated with carrying munitions. Most embarkation, en route, and, perhaps, debarkation bases limit aircraft with munitions to particular portions of the base. The limit on the amount of this ramp space could significantly slow the overall pace of unit deployment—especially for units that expect to engage enemy forces upon arrival.

An Army unit's experience with crisis deployments will also affect the efficiency and speed of aircraft loading. Units that routinely train for specific deployments are far better prepared than those deployed on an ad hoc mission. For example, during the Cold War, annual REFORGER deployments to Europe were more efficient than the Desert Shield deployment, since the personnel knew exactly what was required and how they would be deployed, and had become proficient in all aspects of the deployment.

Because the objective of this work is to broadly understand the important aspects of SBCT deployment given all of the uncertainty, we used airlift planning factors for this analysis.[10] Airlift planning factors were derived from historical averages under different conditions (e.g., payload, aircraft USE rate, amount of Reserve call-up, and on-ground time) to provide the airlift planner with broad measures of aircraft performance. Planning factors can be used to determine the average time and number of sorties required to deploy particular units.

Available Airlift Fleet

The deployability of the SBCT will be evaluated using the airlift projected to be available near the end of this decade. For ease of comparison, the analysis that follows was conducted entirely in C-17 equivalents. The current airlift fleet is in a state of change: C-17s are being introduced, C-141s are being retired, and options for modernizing the C-5 fleet are being discussed. As of December 2001, 80 of the initially programmed 120 C-17s were delivered to the USAF. As of this writing (June 2002), a portion of the funds for an additional 60 aircraft has been appropriated, which would bring the total aircraft

---

[10]HQ USAF, *Air Mobility Planning Factors*, Washington, D.C.: Air Force Pamphlet 10-1403, 1 March 1998, http://afpubs.hq.af.mil/pubfiles/af/10/afpam10-1403/afpam10-1403.pdf.

inventory (TAI) to 180.[11] Delivery of 15 C-17s is expected in 2002. At this rate, the additional 100 aircraft could be produced before the end of 2008. Of the 180 TAI aircraft, about 156 should be Primary Aircraft Authorized (PAA).[12] The retiring C-141s are not considered part of the airlift fleet for this analysis. Two models of the C-5 are currently in the Air Force inventory, totaling 126 TAI aircraft.[13] Of these, about 110 are PAA: C-5As account for 66, and C-5Bs account for the remaining 44. The C-130s are not a good choice for long-range deployments, given their range, speed, and payload limitations. We do not use them in this analysis.

Using the planning factors contained in *Air Mobility Planning Factors*,[14] we calculated that the PAA C-17s and C-5s provide about 239 C-17 equivalents of airlift available to the Air Mobility Command (AMC) daily.[15] Most of the airlift analysis that follows assumes that the SBCT is allocated 60 C-17 equivalents for its deployment—roughly 25 percent of the projected AMC capacity. This value was not chosen to predict the allocation of aircraft to an SBCT deployment but, rather, to provide a reasonable estimate by which to compare the various cases presented in this analysis. It probably is a bit optimistic; during the first two months of Operation Desert Shield, the *entire* Army received only 40 percent of the strategic airlift.[16]

To what degree the airlift fleet will be mobilized and the extent to which civilian carriers will participate (if they do participate) will be a

---

[11]The 180 TAI C-17s does not include the 13 Air Force Special Operations Command (AFSOC) aircraft that are not available to the Air Mobility Command (AMC) on a daily basis.

[12]TAI includes spares and other aircraft not typically configured or available for immediate deployments. For this reason, the PAA figure is used for calculating available aircraft.

[13]This total includes 76 TAI C-5As that are collectively in the Active, Reserve, and Guard inventories and 50 TAI C-5Bs that are currently in the Active force.

[14]HQ USAF, 1998.

[15]This calculation assumes 156 PAA C-17s, 66 PAA C-5As, and 44 PAA C-5Bs; a 45-short-ton payload for the C-17 and a 61.3-short-ton payload for the C-5A/B; and a contingency USE rate of 11.7 for the C-17, 5.8 for the C-5A, and 7.5 for the C-5B.

[16]John Lund et al., *Project AIR FORCE Analysis of the Air War in the Gulf: An Assessment of Strategic Airlift Operational Efficiency*, Santa Monica, Calif.: RAND, R-4269/4-AF, 1993.

national decision. The involvement will depend on the particular crisis and will be affected greatly by the world situation at the time. In a serious crisis, leaders will tolerate substantial disruption to peacetime military airlift operations and the civilian air transport economy. In general, there is less willingness to suffer such disruptions during SSCs. The use of civilian aircraft will also be limited by the security environment abroad; civilian carriers rarely will be asked to fly into threatened airspace or insecure airfields. Throughout this analysis, we assume that no call-up of Reserve forces has been authorized and that no use of civilian aircraft is permitted. Given the limitations of most Third World airports, SBCT deployment timelines are most likely to be constrained by throughput at the final-destination airfield. Thus, adding Civil Reserve Air Fleet (CRAF) aircraft for these contingencies would not speed SBCT deployment, although doing so might be necessary if simultaneous deployments to multiple contingencies are called for.

Airfield Infrastructure

Airfield infrastructure—the number, quality, and lengths of runways and taxiways; the quality and size of ramp space; and the presence of refueling facilities and unloading equipment—determines how rapidly airlifters can land, unload, be serviced, and take off. The aerial port of embarkation (APOE), en route base or bases, and the aerial port of debarkation (APOD) are likely to vary greatly in their capacity. The APOE is likely to be a highly capable airport, because it will be located either in the United States or in a developed country, and the SBCT will usually deploy from this location. If an en route base is used, the airlift planner has some latitude in choosing this base and can choose a base with suitable infrastructure. The APOD is likely to be the least developed base, since the SBCT is envisioned for rapid deployment in contingencies that most commonly occur in less-developed areas. The infrastructure may also have been damaged during the conflict. Finally, since the SBCT will need to be deployed as close as possible to the conflict to avoid a long road march, often only a single APOD will be available.

Our analysis uses throughput[17] as the variable for investigating the effect of airport infrastructure on deployment times. Many factors determine the throughput of a base: parking space for aircraft refueling, maintenance capacity, and the ramp space at the airbase for storing and assembling the SBCT equipment. All of these factors taken together are typically referred to as the *maximum on ground* (MOG)[18] for the aircraft. Here, we use *throughput* as a measure of the number of C-17 loads per hour that can be accommodated. The airlift analysis that follows is conducted for throughputs of 1, 2, and 3 C-17s per hour. For example, a throughput of 2 C-17s per hour means that, for virtually the entire operation, an average of 1 C-17 of cargo must be unloaded every half hour, 24 hours per day, potentially for several weeks (depending on the scenario). We believe this parameter provides a little better indication of the type of infrastructure required than does the amount of ramp space, the resulting number of aircraft that can be parked "cheek-by-jowl" on that ramp, and the time each aircraft needs for unloading.

Deployment of the SBCT from Projected Bases

As stated in Chapter One, the projected bases for the SBCT are located in the United States: Fort Lewis, Washington; Fort Wainwright, Alaska; Fort Polk, Louisiana; and Schofield Barracks, Hawaii. We used 60 C-17s and the following contingency planning factors from AF Pamphlet 10-1403: a 45-short-ton payload or 90 passengers (PAX), an 11.7 contingency USE rate, and 2.25 hours on the ground. Figure 2.1 illustrates the maximum distances that SBCTs based at these locations can reach in 96 hours, represented by the 1,325-nautical-mile (nmi) radius of each of the circles around each of the projected bases. In other words, the SBCT could be deployed to any suitable airport within the circles in this time period.[19]

---

[17] *Throughput* is the number of aircraft that can land, unload, be serviced, and take off per hour.

[18] For a detailed discussion of MOG, see James P. Stucker and Ruth T. Berg, *Understanding Airfield Capacity for Airlift Operations*, Santa Monica, Calif.: RAND, MR-700-AF/OSD, 1998.

[19] According to U.S. Transportation Command (2002, p. 35), during Operation Desert Storm, the minimum average time between launches for the 82nd Airborne Division was an hour, which means that the average throughput for the 82nd Airborne at its de-

RANDMR1606-2.1

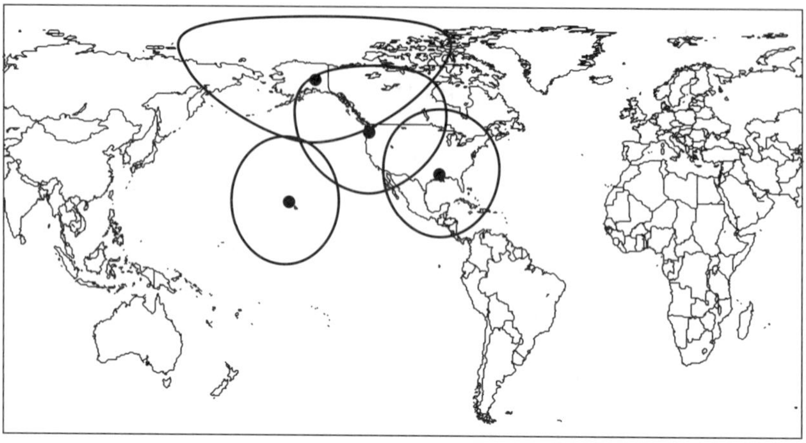

Figure 2.1—SBCT 96-Hour Reach from States of Hawaii, Alaska, Washington, and Louisiana

An important aspect of the deployment—the throughput constraint of the airbases—is not captured in this figure. In order to deploy the SBCT 1,325 nmi in 96 hours, airport throughputs in excess of 4 C-17s per hour would be required around the clock for nearly 4 days. This rate would be a heroic achievement under the best of circumstances—unlikely when considering the quality of airport infrastructure in much of the world.

Figure 2.2 indicates how long it would take to deploy the SBCT from these bases to illustrative locations around the world, using 60 C-17s and a more modest throughput of 2 C-17s per hour (24 hours per day, every day until full unit closure). The fastest deployment is 9 days from Fort Polk to Bogota, Colombia; all other points in South America are reached within 14 days. European scenarios take roughly 14 days. African scenarios require 12 to 21 days. Asian scenarios take from 13 days up. The slowest scenario in the figure is 21 days to Kandahar, Afghanistan. (See Appendix B for more details.)

---

parture airfield (Polk AFB) was 1 aircraft per hour instead of the 4.4 aircraft per hour we postulate for Figure 2.1.

RANDMR1606-2.2

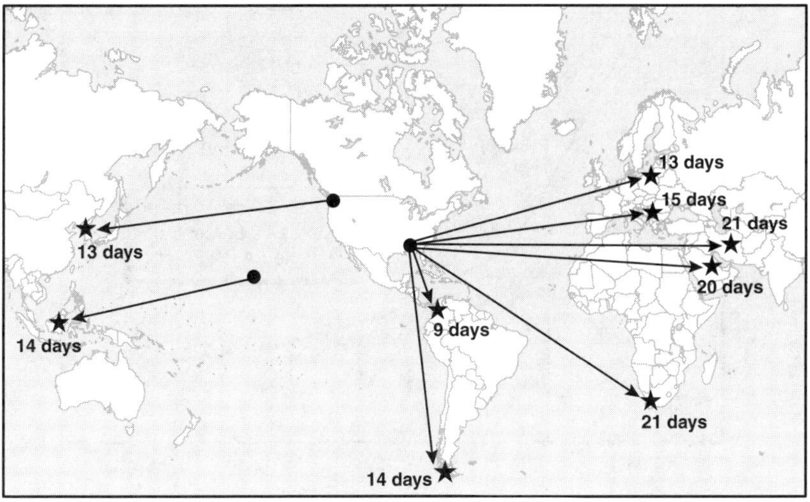

Figure 2.2—Deployment Times by Air for SBCT to Representative Destinations (60 C-17s; Throughput of 2 C-17s per Hour)

It is clear from the preceding discussion that the number of C-17s available to deploy the SBCT, as well as the available throughput of the airports, is critical for determining the number of days a deployment will take. Figure 2.3 presents the relationship between the number of deployment days and these two critical variables for a representative deployment of 5,000 nmi.

The bars in Figure 2.3 are read from the left axis and represent the number of C-17s required to perform the deployment of the SBCT for each of the deployment times listed. For example, a 5,000-nmi deployment of the SBCT in 4 days would require approximately 182 C-17s dedicated to the operation. It is informative to compare this requirement to airlift availability. As discussed earlier, 239 C-17 equivalents of airlift are available to AMC on a daily basis. Conducting this deployment in 4 days would require 76 percent of AMC airlift (assuming contingency-planning payloads and USE rates).

The curve on the chart is read from the right axis and identifies the airport throughput in number of C-17s per hour for each of the de-

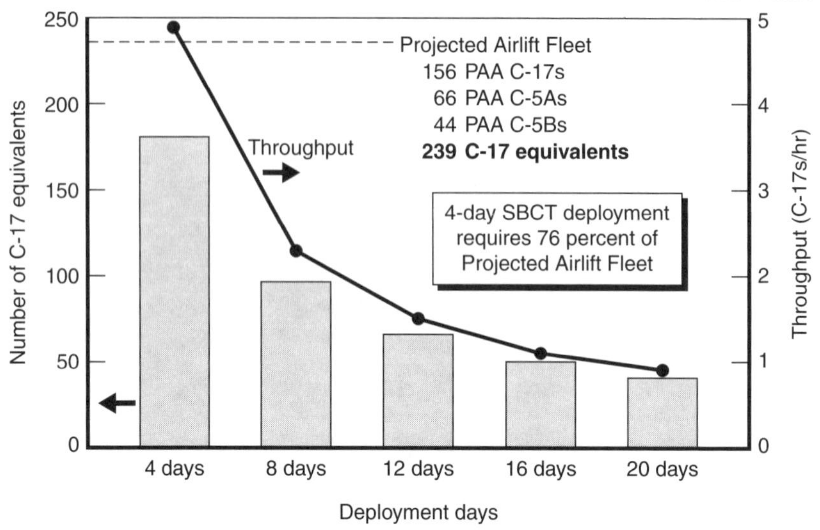

Figure 2.3—Airlift Fleet and Throughput Requirements for a Representative Deployment of the SBCT to 5,000 nmi as a Function of Deployment Days

ployment times (assuming 24-hour operations). It is important to understand that this throughput would be required for nearly the entire deployment time frame. For example, the 4-day deployment requires a throughput of nearly 5 C-17s per hour. Since the first sortie (outbound from APOE to APOD of 5,000 nmi) would take about 12 hours of flight time, this level of throughput would be required for 84 hours (3-1/2 days of 24-hour operations).

A similar chart is presented in Figure 2.4 for a 2,500-nmi deployment. Although the throughput is roughly the same as for the 5,000-nmi deployment, the percentage of the airlift fleet required is much more manageable for this deployment distance. A 4-day deployment can be accomplished with about 42 percent of the projected airlift fleet, but would require almost 5 C-17s per hour. The throughput requirement for the 2,500-nmi case is slightly less than that of the 5,000-nmi case, accounting for the shorter flying time on the first sortie (time that is included in the total deployment time, but not available to unload aircraft).

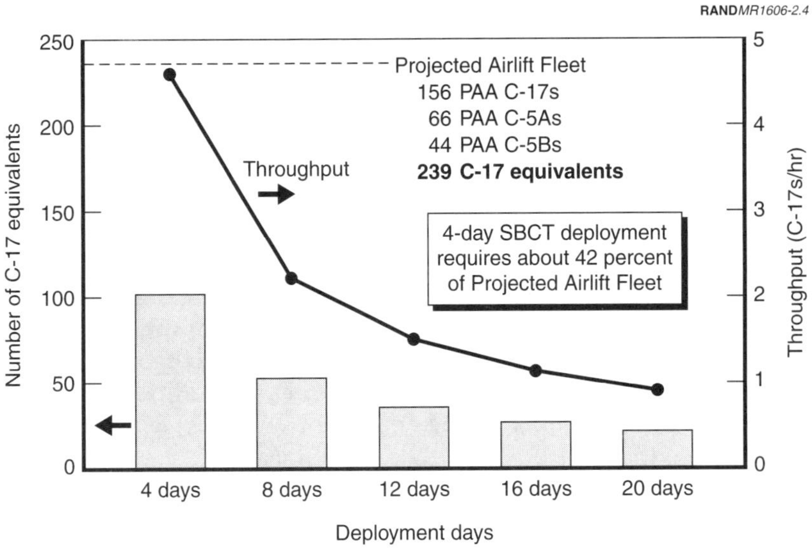

Figure 2.4—Airlift Fleet and Throughput Requirements for Representative Deployment of the SBCT to 2,500 nmi as a Function of Deployment Days

The preceding discussion leads one to ask, What would be required of the deployment bases and the airlift fleet to move the Army closer to achieving its goal of rapid deployability? To answer this question, we first assume that, with careful planning of loads to most efficiently use cargo space, the average payload may be increased. In this example, we chose an average payload as the average of the planning-factor payload and the *allowable cabin load* (ACL), the maximum payload that can be carried on a mission.[20] This would result in an average payload of about 57 short tons for the C-17, including Army personnel flying with vehicles.

Achieving an average payload of 57 short tons is not easy. Although it may be possible to attain heavier payloads on a subset of sorties, other payloads will certainly weigh in significantly lighter. Some sorties may fill all available space (cube out) before they reach maximum payload, and it is unreasonable to assume that the loadmasters will be able to achieve perfect loads for every sortie (even if perfect

[20]The ACL may be limited by the maximum takeoff gross weight, maximum landing gross weight, or by the maximum zero fuel weight (HQ USAF, 1998, p. 23).

loads were possible). This load operation will be further complicated by the quick-deployment timeline. For example, a loadmaster may have space for an additional High-Mobility, Multipurpose, Wheeled Vehicle (HMMWV) on a particular sortie, but loading this vehicle may delay the aircraft departure, thus slowing the entire deployment. Another constraint is that the unit may need to be combat-capable upon arrival at the APOD, which would suboptimize the loading scheme.

The 57-ton average payload is significantly higher than has been achieved historically. Loadmasters would have to optimally load every aircraft and units would have to arrive on time and organize themselves into optimal chalks[21] for loading. In addition, so-called hot loads (ammunition and other dangerous cargoes) will need to be managed so that they do not exceed the airbase limit.

As mentioned earlier in this chapter, Task Force Hawk was deployed to Rinas in Tiranë, Albania, which had an overall MOG of 4 (of which the U.S. military received only half, a MOG of 2). This may very well be what could be expected from much of the Third World. Figure 2.5 presents the days required as a function of MOG. These calculations assume a deployment distance of 2,500 nmi and consider two cases of different average payloads (45 and 55 tons[22]) and two different on-ground times. The first on-ground time considered is the expedited planning-factor time for the C-17 (1 hour and 45 minutes). *Expedited planning-factor time* allows for unload operations only and does not provide time to refuel or reconfigure the aircraft.[23] The second time considered (1 hour and 24 minutes) explores the operational effect of reducing the expedited ground time by 20 percent. Both of these cases would require tankers, significantly increasing the complexity of the operation (mission planning,

---

[21]When a unit deploys by air, it is divided into aircraft loads, called *chalks,* each assigned to a particular aircraft.

[22]Carrying troops and their personal gear with the vehicles would increase each payload by about 2 short tons.

[23]HQ USAF, 1998, Table 5, Ground Times, p. 15.

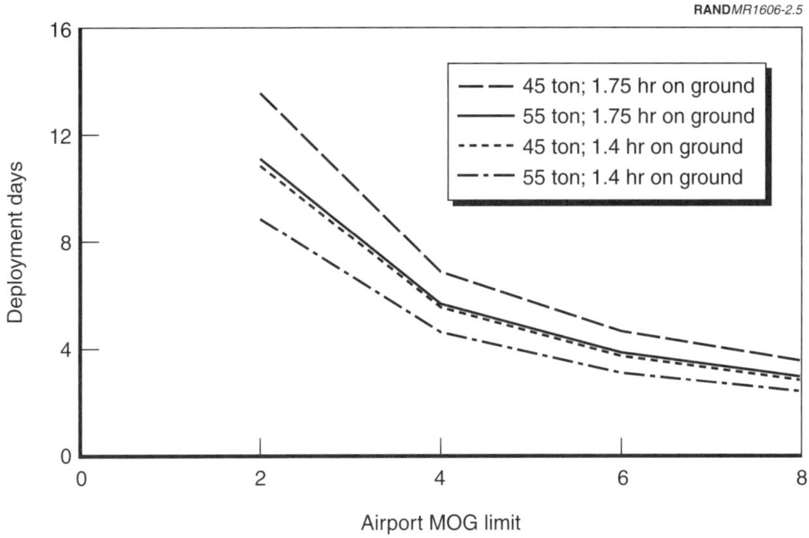

Figure 2.5—Effect of Airbase MOG on Deployment Days, Using Alternative C-17 Loading and On-Ground Times

allocation of scarce tankers, additional host-nation bases for the tankers, etc.).

As can be seen from the figure, optimal loading of aircraft and reduced on-ground times can reduce the MOG requirement. For example, if we can achieve 55-ton average payloads, reduce ground time to 1.4 hours, and relax the time constraint somewhat to a week, the MOG requirement drops from over 6 to 3, a much more realistic assumption for operations in the developing world. If we further relax the time constraint to 8–9 days, the MOG requirement drops to 2.

To accomplish high-tempo mobility operations at bases with limited infrastructure, the airlifters and SBCT will be required to operate in a highly efficient manner. The SBCT will need to quickly move from the aircraft and assemble at a location that does not impede airlift operations. The USAF ground personnel will need to move aircraft in and out quickly and efficiently with limited queuing and traffic delays. Maintenance personnel must quickly diagnose problems, fix them, and get the aircraft airborne to the APOE, if possible. Extra ramp space and towing gear will likely be needed for aircraft with more-serious mechanical problems. SBCT personnel will also need to be aware of the potential for foreign object damage (FOD) to aircraft, limiting activity that could kick debris onto the runways, taxiways, and ramps—probably not easy to avoid, given the poor condition of airports likely to be encountered during SBCT deployments. A final point on the MOG issue is that these short ground times at the APOD may actually increase ground times at the APOE, since maintenance actions that did not prohibit the aircraft from getting airborne at the APOD will need to be addressed at the APOE.

Not considered in this analysis is the possibility that the SBCT will get heavier. The Stryker IAVs in the SBCT have limited armor protection, and for some contingencies may need to deploy with bolt-on reactive armor for each vehicle.[24] Also, the level of sophistication of all vehicles in the SBCT is proposed to be very high, and these high-tech vehicles may be heavier than their stock counterparts. Given the typical trend of military acquisition to add more subsystems to correct unforeseen deficiencies, their weight may increase as the unit evaluates performance and identifies deficiencies. Finally, the SBCT that this analysis considers does not include divisional or corps-level assets (e.g., helicopters), which may be required for some operations.

SEALIFTING THE SBCT

The stringent constraints—primarily on weight—imposed by the requirement to be air-transportable are relaxed when the mode of transportation is sealift. However, the use of sealift to deploy the SBCT presents an entirely new set of issues that must be considered. This section presents our analysis of deployment of the SBCT by sealift, using U.S. DoD sealift planning factors.

---

[24]The potential requirement for reactive armor was a topic of the 311th Support Command (Corps) Joint Operations Conference, Universal City, CA, 15–17 March 2002, which one of the authors [DTO] attended. An initial estimate on the deployment weight of these reactive kits is 3.75 short tons per vehicle. Given that the SBCT has 327 Strykers, deploying enough reactive armor just for these vehicles would increase the deployment weight of the unit equipment by nearly 10 percent (an increase of over 1,225 short tons).

If good ports are available and the area of operations is within a couple days' road march of the port, sealift is surprisingly competitive with airlift. Figure 2.6 illustrates this similarity for most of the destinations shown in Figure 2.2, using two Fast Sealift Ships transiting at 27 knots (kt). Adding the road march, we see that sealift takes 11 days to get to Bogota, Colombia, rather than 9 by air. Indonesia takes 14 days by air or by sea. Sealift takes only 16 days to Cape Town, South Africa, instead of 21 days by air. Air and sea are essentially the same for our European and Saudi deployments. For Afghanistan, the long road march makes sealift slower at 27 days versus 21 by air. However, as we show in the following discussion, good ports are difficult to find in the parts of the world where the SBCT is most likely to be deployed. For deployments to poorer ports, long-range sealift using large, deep-draft ships is less attractive.

RANDMR1606-2.6

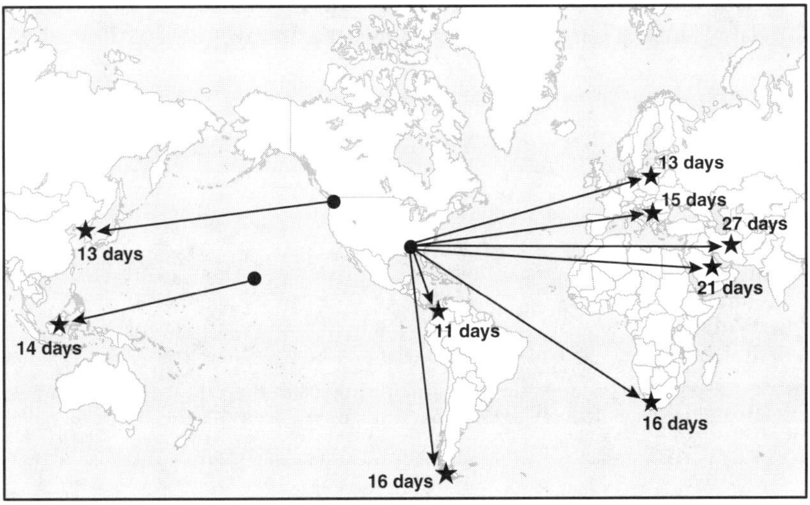

Figure 2.6—Deployment Times for Sealift Using Two Fast Sealift Ships at 27-kt Transit Speed

Sealift Assets Considered

Our sealift analysis was conducted using roll-on/roll-off (RoRo) ships. The Fast Sealift Ships (FSS) and the Large, Medium Speed RoRo (LMSR) are the high-speed ships that the Military Sealift Command (MSC) uses for deployment from CONUS. These ships are very long range (12,000 nmi) and are kept in a reduced operational status of 4 days (ROS-4), which means that the ships require 4 days to activate and prepare for deployment operations. Our analysis assumes that, in these 4 days, the deploying units prepare to deploy, move to the port, and begin to load as soon as the ships are available.[25] MSC has eight FSS and seven LMSR in its inventory, all located on the Atlantic and Gulf coasts of the United States. Table 2.2 presents the physical characteristics and capabilities of these vessels.

We first determined the number of ships required to deploy the SBCT.[26] During Operation Desert Storm, the average loading of an

Table 2.2

Characteristics of FSS and LMSR Ships

Ship Characteristics	FSS	LMSR
Speed (kt)	27 sustained; 33 max	24
Displacement (long tons)	55,350	55,300–65,000
Length (ft)	946	907–954
Beam (ft)	105.5–106	105.5–106
Draft (ft)	34.8–36.9	34–35.75

SOURCE: These data were compiled on 17 July 2001 from the Military Sealift Command website, http://www.msc.navy.mil, on the various ships in the inventory.

---

[25]As discussed above, the "deployment clock" starts ticking when the ships begin to load—after the 4 days required by ROS-4.

[26]Our analysis assumed that the equipment would go by sea and the personnel by air, meeting at the SPOD. In many cases, the deployment of personnel could be done with a few chartered flights. For example, using Boeing 747s to deploy the personnel would require about nine sorties (each 747-400 can carry about 500 passengers, depending on seating configuration).

FSS was 850 vehicles.[27] The SBCT consists of about 1,145 vehicles: 327 IAVs and 818 other vehicles. Based on the Desert Storm average load, two FSSs would be required to transport the SBCT. It may be possible to transport the entire SBCT with one LMSR, which is somewhat larger than the FSS, but that remains to be determined. In any event, using two vessels will decrease the unload time, if both ships can be unloaded simultaneously.

Calculating Sealift Deployment Time

Three main functions are required for SBCT sealift deployment (port-to-port): load, sail (or steam), and unload. Computation of the sail portion of the deployment time is fairly straightforward and is simply the deployment distance divided by the speed of the vessel. For the calculations that follow, 27 kt is used for the FSS and 24 kt for the LMSR.[28]

Much as for airlift, the time to load and unload depends on the port infrastructure. The FSS and LMSR are very large ships. Length and draft limitations of many ports make it impossible to support them. Our base case assumes that the ships are in a port or harbor capable of supporting simultaneous unloading. A discussion of unloading operations when the harbor cannot accommodate vessels of this size is presented in the next subsection.

To better understand the time required to unload RoRo ships, we consulted two historical cases: Desert Shield and REFORGER. In Desert Shield, an average of 15 to 16 vehicles per hour were unloaded; REFORGER exercises averaged 24 vehicles per hour for loading and 35 vehicles per hour for unloading.[29] The Desert Shield and REFORGER data represent a fairly wide range, and neither set is quite right for our purposes. For Desert Shield, on the one hand,

---

[27]See Headquarters, Department of the Army, *Transportation Reference Data,* Washington, D.C.: Field Manual 55-15, October 1997, Table 5.2.

[28]Weather, which could slow the deployment, is not considered here. Bad weather will likely result in all ships (regardless of type) slowing to about 15 kt. Weather can also slow airlift operations, although an aircraft's speed gives aircraft more options to route around bad weather. For consistency, our analysis assumed good weather for both airlift and sealift.

[29]Headquarters, Department of the Army, 1997, Tables 5.2 and 5.4.

many of the vehicles deployed were tracked and the units were not designed for strategic mobility, as is the SBCT. On the other hand, REFORGER, an annual practice of a massive movement of U.S. forces to well-prepared ports in Europe, probably overstates what could be accomplished during a more ad hoc deployment to a new location. Every aspect of the movement was carefully planned, elaborate procedures were developed, and great expertise was acquired through years of practice. It seems likely that SBCT load/unload times would be slower than the REFORGER experience but faster than Desert Shield. For these reasons, we chose a value in the middle of the range: 20 vehicles per hour for both load and unload.

Since our base case assumes two ships simultaneously loading and unloading, computation of the total time is a simple division: 572 vehicles per ship divided by 20 vehicles per hour equals 28.6 hours to load or unload the ships. In addition to this load/unload time, one-half day (12 hours) is assumed at each port to account for port time, docking (or cast off), tides, and weather.[30] This additional one-half day results in a total time of 40.6 hours for both the loading and unloading operations.

As stated earlier, many of the world's ports are incapable of accommodating ships of this size. For example, the SBCT would be an excellent force for peace operations in Africa. Yet, few African ports can accommodate FSS or LMSR ships.[31] Thus, if sealift of the SBCT is deemed desirable, a means must be found to give sealift more of an expeditionary capability.

---

[30]Although Field Manual 55-15 (Headquarters, Department of the Army, 1997) states that an additional day of port time should be added to the estimates of load/unload time, our analysis added a half-day to the time for both load and unload, since this was the portion of time assumed to be within the "port-to-port" ground rules.

[31]Our initial assessment indicates that Egypt, Algeria, Senegal, and South Africa have at least one port capable of accommodating ships of this size. This assessment is based on the information contained in National Imagery and Mapping Agency (NIMA), *World Port Index,* Pub 150, http://164.214.12.145/pubs/pubs_j_wpi_sections.html, November 1, 2002. Our criteria for ports to qualify as capable of handling LMSR or FSS ships are as follows: (1) The entrance channel and harbor by the cargo pier are 36 feet or greater in depth; (2) the port was capable of handling *large ships,* defined as over 500 feet long in Pub 150, and (3) Pub 150 classified it as a "Large"- or "Medium"-class harbor.

Outside Harbor Unload Operations

The SBCT is designed for SSCs and will likely be deployed to countries with poor (or perhaps damaged) infrastructure. Therefore, in the absence of a harbor capable of accommodating large MSC vessels to unload the SBCT, the Navy can use a roll-on/roll-off discharge facility (RRDF) and lighterage to bring the vehicles ashore—either to the beach or to a harbor.

The RRDF is a floating pier that is set up off the coast of the deployment location. Vehicles and cargo from the large oceangoing vessels are transloaded at the RRDF to smaller vessels and taken ashore. In this section, we use DoD planning factors to evaluate (1) RRDF assembly and preparation for operations and (2) transloading the SBCT vehicles and moving them to shore with lighterage. Figure 2.7 illustrates the basic RRDF components.

RRDF Assembly. The planning and reference documents do not provide a consistent estimate of the RRDF assembly time. Using the USMC *MEF Planner's Reference Manual,*[32] we arrived at a total time of 44 to 48 hours, whereas we arrived at 16 to 22 hours using the information in *Joint Tactics, Techniques, and Procedures for Joint Logistics Over-the-Shore (JLOTS).*[33] Table 2.3 presents the components of this estimate for each of the two publications.

For our analysis, we chose an estimate in the middle of the range, 1.5 days, to assemble the RRDF and prepare for operations. A question that remains unanswered by this research is whether the RRDF would be assembled before the arrival of the sealift ships. If the RRDF could be assembled before the SBCT arrived, unloading would take about a day less. Our analysis assumes that the RRDF equipment comes with the sealift ships carrying the SBCT.

---

[32]USMC, Quantico, Va.: MAGTAF Staff Training Program, MSTP Pamphlet 5-0.3, 1999.

[33]Joint Chiefs of Staff, Washington, D.C.: Joint Publication 4-01.6, 12 November 1998.

RANDMR1606-2.7

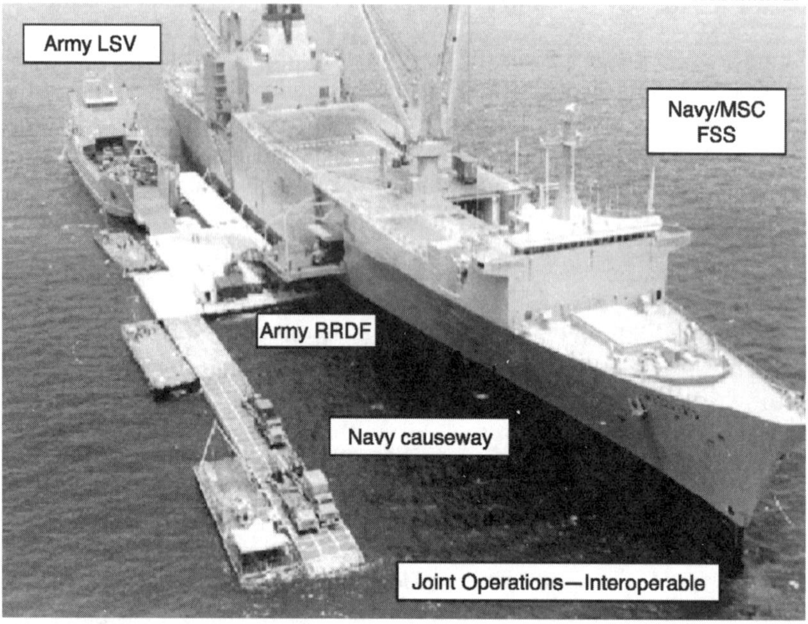

Photo courtesy of U.S. Transportation Command.
See http://www.jdtc.transcom.mil/images/Addnl_JLOTS_images/additional_jlots_imagery.htm.

Figure 2.7—Unloading Operation Using RRDF and Lighterage

Once the RRDF is assembled, the task of unloading the vehicles can begin. We again used planning factors to estimate the time required to unload the vehicles from the ships to shore. We assumed that the sealift ships and the RRDF would be anchored about 1 nmi off the coast of a shallow harbor that could accommodate lighters. We used the LCU-2000 as the representative lighter to take the vehicles from the RRDF to the shallow harbor. Table 2.4 presents the specifications of the LCU-2000, a fairly large lighter capable of carrying 13 wheeled vehicles per sortie.

Lighterage. Our planning-factor timeline analysis for the LCU-2000 lighterage operation is presented in Table 2.5.

Table 2.3

Components of RRDF Assembly for Two Documents

Source / Function	Time (hr)
MSTP Pamphlet 5-0.3 (USMC, 1999)	
Offload side-loadable warping tug (SLWT) and LCM8s (Landing Craft Mechanized)	1.5
Moor anchors	6.0
Position ship's ramp	0.5
Assemble RRDF	36–40
Total MSTP Pamphlet 5-0.3 planning time	44–48
Joint Pub 4-01.6 (JCS, 1998)	
Offload lighterage	10
Assemble RRDF (self-sustaining/non–self-sustaining)	6–8/12
Total Joint Pub 4-01.6 planning time	16–22

Table 2.4

Specifications of LCU-2000

Length	175 feet
Beam	42 feet
Draft (full load)	
Forward	5 feet
Aft	9 feet
Cargo capacity	350 short tons
Light load	10,000 nmi @ 12 kt
Full load	6,500 nmi @ 10 kt

SOURCE: Joint Chiefs of Staff, *Joint Tactics, Techniques, and Procedures for Joint Logistics Over-the-Shore (JLOTS)*, Washington, D.C.: Joint Publication 4-01.6, 12 November 1998, Figure B-1.

To minimize the ship unloading time for this analysis, we first assumed that enough LCU-2000s are available so that the lighter stations are never vacant. Therefore, the limiting factor in unloading the ship is the amount of time it takes to move the vehicles from the

Table 2.5

LCU-2000 Lighterage Operation Planning Factors

LCU-2000 Function	Time (minutes)
Approach and moor	9
Load	96
Cast off and clear	4
Steam (1 nmi to shore)	8
Approach and moor	12
Unload	30
Cast off and clear	7
Steam (1 nmi to RRDF)	8

RRDF to the LCU-2000s and cast off.[34] We assumed that two sealift ships are being unloaded on two RRDFs simultaneously. We further assumed that each RRDF has two unloading stations, so that each RRDF can accommodate 2 LCU-2000s simultaneously (a total of 4 LCU-2000 stations). Thirteen vehicles can be transloaded from the RRDF to an LCU-2000 at each of the unloading stations. The time that the lighter station is "tied up" to unload these 13 vehicles is the time it takes the LCU-2000 to approach and moor, load, cast off, and clear: 1 hour and 46 minutes. Therefore, 26 vehicles can be moved every hour and 46 minutes. The total operation consisting of 1,145 vehicles of the SBCT being unloaded from two sealift ships, using two RRDFs (one for each ship), would take about 39 hours. Adding to this amount the 1.5 days (36 hours) calculated earlier to assemble the RRDFs and prepare for operations, the entire operation would take about 3.1 days. Figure 2.8 illustrates a notional lighterage operation from RRDFs to a port.

The number of LCU-2000s required for such an operation turns out to be fairly small. Given the planning factors identified above, the

[34]The factors in Table 2.5 were taken from Joint Chiefs of Staff, 1998, Table A-2, Discharge Planning Factors. Since port unload times were not available in this source document, the times presented here were for the case for unloading on a floating causeway, which may be slightly different from those at a port. Because the analysis focuses on the unload rates at the RRDF as the limiting factor, these differences will not change the overall time. In the unlikely event that the times are significantly different, more lighters could be added to the operation so that the unloading at the RRDF is held as the limiting factor.

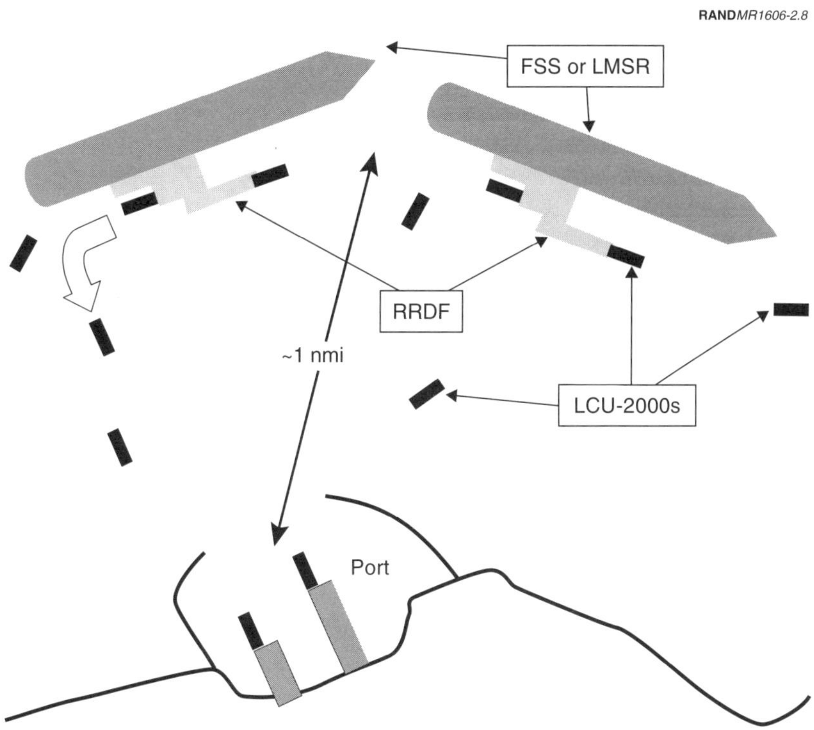

Figure 2.8—Notional Lighterage Operations from RRDF to Port

time that an LCU-2000 spends at the RRDF is 1 hour and 46 minutes. All other times (round-trip steam, unload, etc.) account for only 1 hour and 5 minutes. Therefore, only two operational LCU-2000s would be required per unloading station to ensure that an RRDF unloading station is never waiting for a lighter to arrive.

Unloading operations using the RRDF and lighterage are feasible and reasonably fast in gentle sea states, to a maximum sea state of 2 on the Pierson-Moskowitz Sea Spectrum, but are severely limited in poor weather. For JLOTS operations, sea state 2 is defined by wave heights from 1.5 to 3.0 feet and wind speeds of 5 to 12.7 kt. A sea state of 3 is defined as wave heights of 3.5 to 5.0 feet and wind speeds of 13.7 to 16.4 kt. Weather conditions of this magnitude are common in many regions of the world throughout the year. Joint Pub 4-01.6 for JLOTS cautions against operations in poor weather:

> The inherent risks of operation in sea state 3 are not worth the minimal productivity and possible equipment damage which could occur. Such damage could prevent timely resumption of operations as weather clears.[35]

These weather concerns and the time delays associated with lighterage operations indicate that sealift ships capable of unloading directly into a harbor are the most desirable.

Shallow-Draft Sealift

An additional option for sealifting the SBCT is the high-speed catamaran ("fastcat") ferry, a vessel that typically has drafts on the order of 10 to 15 ft and lengths up to 120 meters (m) and can, therefore, access smaller ports. Although catamarans are primarily used for civilian service to ferry passengers and automobiles, their military applications were pioneered by the Australians to support UN operations in East Timor.

The Royal Australian Navy uses the *Jervis Bay,* an 86-m, 1,250-ton, 40-kt wave-piercing catamaran to make the 430-mile run (up to three times weekly) between Darwin, Australia, and Dili, Indonesia. The commander of the *Jervis Bay* describes their first run into Dili at the beginning of UN efforts to stop the violence in East Timor:

> The first time we made a run into Dili, the port was totally trashed. There was a lot of confusion, a lot of things strewn on the wharf and there were no port services. With nothing to help us, we were still able to land troops quickly. The catamaran definitely gave us a big advantage.[36]

A somewhat larger vessel, called the *Westpac Express,* is being leased by the USMC as a theater support vessel for operations between

---

[35]Joint Chiefs of Staff, 1998, Section VII-6.

[36]William Polson, "Navy Goes Down Under, Explores Future of Amphibious Warfare," http://www.c7f.navy.milnews/2000/09/16/html, posted September 16, 2000; website last accessed June 24, 2002.

Okinawa and mainland Japan. It carries 100 HMMWVs, four trucks, 12 UH-1N utilty helicopters, and 950 fully equipped Marines.[37] A third vessel is being tested by the U.S. Army, Coast Guard, and Navy under the *Joint Venture* High Speed Vessel program. The *Joint Venture* is a 96-m catamaran built by INCAT at Hobart, Tasmania (see Figure 2.9).[38] In previous trials, it achieved a top speed of 50 kt (93 km/hr). This vessel will be tested on its ability to carry 450 tons of cargo and 325 personnel to a range of 1,100 nmi at an average speed of 35 kt and to launch and recover helicopters, all in sea state 3. For these tests, it will be fitted with a Navy helicopter deck and vehicle ramps to accommodate tracked vehicles.

The *Joint Venture* and ships of its class do not have the range or payload to be attractive for long-range deployments, but they offer a promising option for regional deployments. For example, if the SBCT were based abroad or had some or all of its equipment prepositioned forward,[39] fast catamarans would enable rapid deployment out to distances of 1,000 to 2,000 miles (refueling would be required for operations much beyond 1,000 miles).[40] With unit sets in Panama, Darwin, and Diego Garcia, the SBCT could cover the Caribbean, Southeast Asia, and Indian Ocean littorals, many requiring no refueling or one refueling at most. Deployment of the SBCT using the 450-ton cargo figure would require about 34 *Joint Venture*

---

[37]Ian Bostock, "USMC Charters High-Speed Catamaran," *Jane's Defence Weekly*, 11 July 2001, p. 3.

[38]The *Jervis Bay* also was built by INCAT. The *Westpac Express* was built by Austal, another Australian shipyard.

[39]In this analysis, we assumed that prepositioning would mean that an entire unit set of equipment plus supplies would be located forward, either on board ships or on land. Our RAND colleagues John Halliday and Eric Peltz note, however, that tactical wheeled vehicles such as HMMWVs account for only 10 percent of the cost of an SBCT but account for roughly 71 percent of the airlift sorties. The most cost-effective strategy, therefore, might be to forward-base these vehicles and supplies and use airlift to move the 300 or so expensive IAVs. This and other mixed strategies deserve further exploration.

[40]New refueling concepts would need to be developed for fast sealift operations. Traditional oilers lack the speed to keep up with the fastcats. One possibility would be to have them routinely at sea or in port along likely deployment routes, which seems impractical or at least expensive at first blush. Another possibility would be to use some catamarans as refueling ships.

RANDMR1606-2.9

Photo courtesy of INCAT. Reprinted with permission.

Figure 2.9—The *Joint Venture,* a High-Speed Catamaran Ferry Built by INCAT at Hobart, Tasmania

sorties.[41] Alternatively, catamarans could be combined with more-traditional sealift and airlift to offer the ability to tailor a deployment to meet specific operational requirements. For example, a smaller fleet of catamarans might be acquired to ensure that an SBCT battalion could access any port, with follow-on forces using LMSRs and lighterage or airlift to move to the operational area.

The utility of shallow-draft vessels in specific scenarios is analyzed later in this chapter.

---

[41]See http://www.incat.com.au/news/news.cgi?articleID=63122. Website last accessed January 15, 2002.

Road March

After the forces are either airlifted or sealifted, a road march may be required. We discuss road marches in this section because an extended road march is more likely for sealifted forces than for those who have been airlifted (an airport is more likely to be within proximity to the operational location than a seaport). Note that sealift distances are given in nautical miles and that road-march distances are in statute miles.[42]

Ironically, we were unable to find U.S. Army planning factors for road marches. We were, however, able to find an Army manual that estimates enemy road-march speeds. The Army predicts that enemy wheeled forces should cover 240 to 480 km per day for dry pavement, 180 to 300 km per day for dry dirt roads, and 80 to 180 km per day for muddy, hilly, and/or urban roads—estimates that vary by a factor of 6. In addition, a march of over 1,000 km requires a day for rest and maintenance.[43]

Estimating the road-march time for the SBCT using these planning factors is not particularly easy. It is likely that the SBCT will be more proficient at road march than the enemy described in the field manual. However the SBCT will find Third World roads inferior to the European roads assumed in the Army analysis and will probably not cover as much as the upper-end estimates. Further, the SBCT operation will not have received anywhere near the planning or intelligence preparation that could be expected from a developed country invading a neighbor. Finally, the planning-factor estimates are for administrative marches, but the SBCT may come under enemy fire during the road march. Therefore, in view of the uncertainty regarding the road types that the SBCT may encounter, we chose an estimate in the middle of the entire 80–480-km-per-day range of 280 km per day, or about 170 miles per day.

In addition to the actual time on the road march, time is required to prepare for the march: for marshalling and route reconnaissance, and for forming up units and planning the road march. All of these

---

[42] 1 nautical mile = 1.151 statute miles.

[43] Headquarters, Department of the Army, *Armor and Mechanized-Based Opposing Force Operational Art*, Washington, D.C.: Field Manual 100-61, 26 January 1998.

functions are assumed to require 1 day, which is referred to as *assembly time* later in this chapter.

DEPLOYMENT TIMES

Now that we have examined differences in deployment methods, we look at deployment times for each method from different locations.

Representative Scenarios

To explore the implications of a wide range of assumptions regarding location of contingency, use of preposition sites, road-march distances, and general operational characteristics of the deployments, in this section we present three representative scenarios comparing airlift and sealift deployments and estimates of the time required to deploy one SBCT to each scenario.[44] First, the deployment is made from CONUS. Then, the deployment time for each scenario is calculated from various preposition sites around the world: airlift, 20-kt sealift (LMSR), a notional 40-kt deep-draft ship, and a 40-kt, shallow-draft ship like the catamaran ferries. Our first scenario has Kosovo as its destination.

Kosovo. This scenario considers the deployment of an SBCT to Skopje, Macedonia, for offensive operations against Serbian forces in Kosovo. In this scenario, the forces are assumed to assemble at the Skopje airport for operations within Kosovo. For the CONUS airlift case, the forces are deployed from Fort Polk[45] in Louisiana, fly 3,250 nmi from Fort Polk to Lajes AFB, in the Azores, off the coast of Portugal, then 2,200 nmi from Lajes AFB to Skopje airport.

For the CONUS sealift deployment, departure is from Beaumont, Texas, across the Atlantic Ocean, through the Strait of Gibraltar and the Mediterranean Sea, then into Thessaloniki, Greece—a journey of approximately 6,378 nmi. Both Beaumont and Thessaloniki are first-

[44]In total, we looked at 10 deployment cases. See Appendix B for details.

[45]Fort Polk does not currently have an airfield suitable for strategic airlift operations. If such operations are to be launched from there, either the airfield would need to be expanded or the brigade would have to road-march to a large airfield, such as Barksdale AFB, Louisiana, or a civilian airport.

rate ports that have the capability to accommodate FSS or LMSR ships. Once the ships are offloaded at Thessaloniki, a road march of 136 miles is required.

The preposition locations assumed for this scenario are Ramstein Air Base (AB) in Germany for the airlift case and Camp Darby, Italy, for the sealift case.[46] The air distance from Ramstein AB to Skopje is 700 nmi; the steaming distance from Camp Darby to Thessaloniki is about 1,100 nmi. Once again, a road march of 136 nmi from Thessaloniki to Skopje is required. (See Figure 2.10.)

Rwanda. This scenario considers the deployment of an SBCT to Kigali, Rwanda. For the CONUS airlift case, the forces are deployed from Fort Polk in Louisiana and flown directly to Kigali, a distance of about 7,100 nmi, requiring flight times in excess of 16 hours. We assumed that tankers would be available to conduct this mission and that overflight of several African countries would be acceptable.

This scenario is presented to show the benefit of direct airlift flights, although such flights are not standard USAF operational practice.[47] The C-17s that would be deployed are capable of aerial refueling, a procedure not usually part of deployment operations. First, tankers are a limited resource. Other missions (e.g., deploying the AEF) are higher-priority claimants to these assets. Second, an en route stop allows for crew change, making sustained, high-tempo, safe airlift

---

[46]The 2001 *Quadrennial Defense Review Report* "envisages that an IBCT should be stationed in the European area by 2007" but is not clear on the specifics. In our analysis for the Kosovo scenario, we considered two possibilities: (1) the brigade is stationed in Germany (notionally, in the Saarbrücken area near Ramstein AB) and air deploys to most contingencies and (2) a brigade set is prepositioned at Camp Darby in Italy, giving the option of either sea deployment from the port at Livorno or air deployment from the airfield at Pisa. For the sea-air comparison presented later in this chapter, we use Darby for the sea deployment and Ramstein for the air deployment. Ramstein is farther from Skopje than is Camp Darby (roughly 800 nmi versus 500 nmi), which, theoretically, disadvantages air in our comparisons—but not in reality. The limiting factor is not the flight time but the MOG at Skopje. For operations that use in excess of about 20 C-17 equivalents and given a MOG limit at Skopje of about 4.5, the deployment will require about 8.67 days, regardless of which of the two locations is used. See Donald Rumsfeld, *Quadrennial Defense Review Report,* Washington, D.C.: U.S. Department of Defense, 30 September 2001, p. 20.

[47]Table B.1 in Appendix B shows deployment data for a Rwanda scenario using in-transit bases instead of aerial refueling.

RANDMR1606-2.10

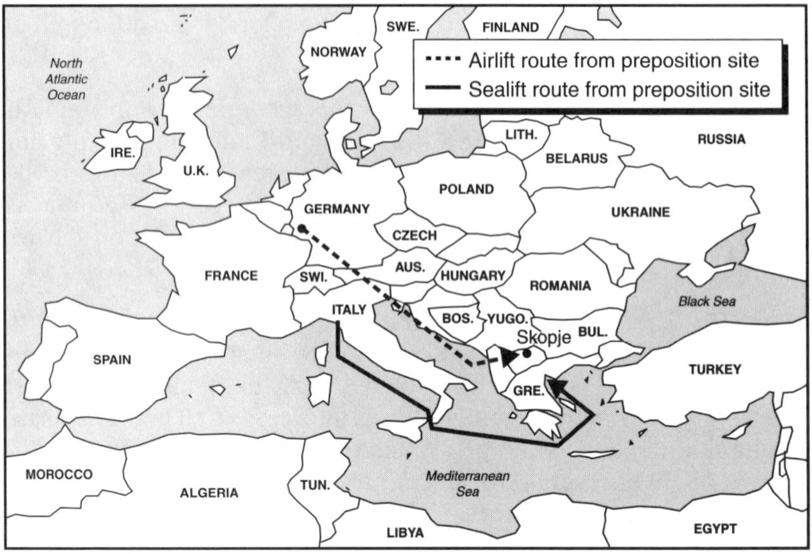

Figure 2.10—Air and Sea Deployments, Germany to Skopje and Italy to Skopje

operations more likely. Third, mission planning is much more complicated when tankers are involved. Most important, aerial refueling only rarely speeds deployments, owing to MOG constraints at destination airfields.[48]

For the CONUS sealift deployment, departure is from Beaumont, Texas, across the Atlantic Ocean, through the Strait of Gibraltar and across the Mediterranean Sea, through the Suez Canal, along the east coast of Africa, and into the port of Mombasa, Kenya—a journey of approximately 9,200 nmi. Mombasa is a port that *may* be capable of accommodating FSS and LMSR ships. For this analysis, we assumed that lighterage would be used to conduct the final deployment of the SBCT into Mombasa. Once the vessels are offloaded at Mombasa, a very long road march of 930 miles is required.

---

[48]The U.S. Transportation Command study (2002) found that air-to-air refueling speeded deployment of the SBCT in only two of 10 scenarios analyzed.

Diego Garcia, an island in the Indian Ocean and the preposition location assumed for this scenario, has an air distance to Kigali of 2,600 nmi; the steaming distance from Diego Garcia to Mombasa is about 1,900 nmi. Once again, a road march of 930 miles from Mombasa to Kigali was required. (See Figure 2.11.)

Indonesia. In the final scenario, we considered deployment of an SBCT to Indonesia—specifically, to the East Kalimantan region on the island of Borneo for an unopposed peace operation. The port city of Tanjung Bara (near Samarinda) was chosen as the deployment location for both the sealift and airlift cases. For the airlift case, the forces were deployed from Schofield Barracks in Hawaii (Hickam AFB) and flown to Andersen AFB on Guam for a refueling stop—3,300 nmi—then on to Tanjung Bara airport—about 1,800 nmi.

RAND*MR1606-2.11*

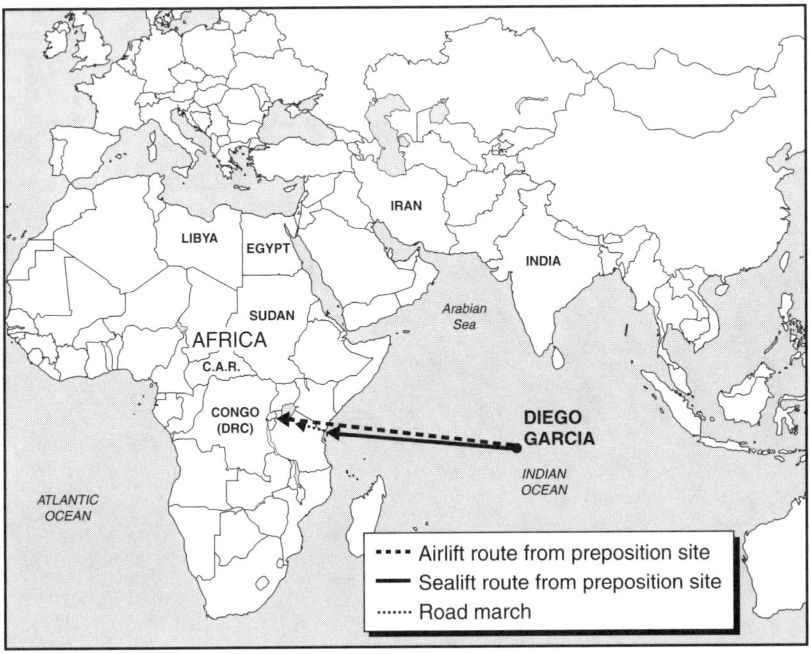

Figure 2.11—Air and Sea Deployments, Diego Garcia to Rwanda

For the sealift deployment, departure was from Pearl Harbor, across the Pacific Ocean, to the port of Tanjung Bara—a steaming distance of about 5,100 nmi. Lighterage was necessary to access this port. Because this port was the final destination for the SBCT, no road march was required.

The preposition location for this scenario was assumed to be the island of Guam. The air and sea distances from Guam to Tanjung Bara are about 1,800 nmi each. (See Figure 2.12.)

Airlift Deployment Analysis

The deployment locations and critical distances discussed in the preceding subsection for each scenario are summarized in Table 2.6. In all scenarios, the contingency planning factors used were 45 short tons, or 90 PAX, and an 11.7 contingency USE rate.

For each scenario, two operational factors play a critical role in determining the number of days required to deploy the SBCT: the number of C-17s allocated to the operation and the throughput of

RANDMR1606-2.12

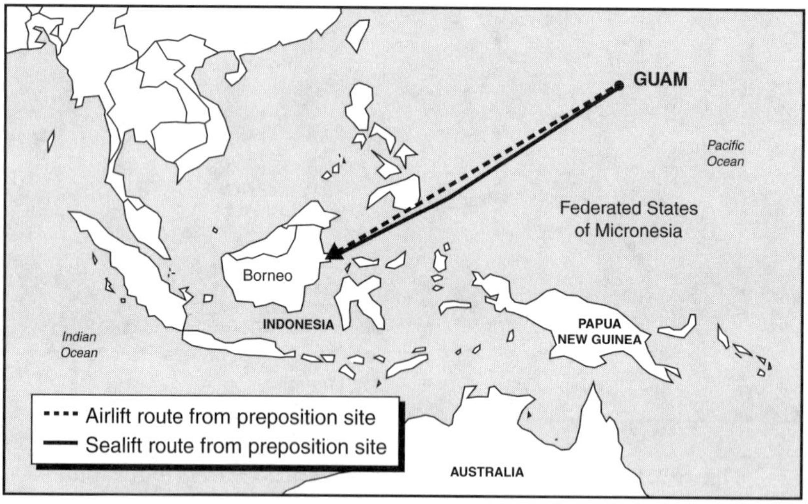

Figure 2.12—Air and Sea Deployments, Guam to Indonesia

Table 2.6

Critical Scenario Information: Airlift

	By Air				
	From the United States			From Regional Prepo Site	
Deployment Location	APOE	En Route Base	Distance (nmi)	APOE	Distance (nmi)
Skopje, Macedonia	Fort Polk	Lajes Field	3,250 + 2,200	Ramstein AB	700
Kigali, Rwanda	Fort Polk	(Aerial-refueled)	7,100	Diego Garcia	2,600
Tanjung Bara, Kalimantan	Hickam AFB	Andersen AFB	3,300 + 1,800	Anderson AFB	1,800

the airports. Figure 2.13 presents the time required to deploy the SBCT from the United States as a function of these two factors for each of the scenarios considered.

Figure 2.14 presents a similar analysis of deployments from the preposition sites. The lines on the charts show the throughput constraint in terms of C-17s per hour. That is, the number of C-17 sorties per hour, 24 hours per day, for the entire operation.[49] The x-axis measures the number of C-17s allocated to the operation. As the reader can see from Figure 2.13 and some scenarios in Figure 2.14, as more C-17s are added to the operation, the curves flatten, indicating that the throughput constraint has been reached. For the shorter-range case presented in Figure 2.13—Kosovo—the throughput constraint dominates; the number of airlifters has no bearing on the deployment time. In short, good throughput and a larger airlift fleet drive CONUS deployments by air to under 10 days.

[49]That level of C-17 throughput is required for the number of deployment days minus the time required to fly the first sortie, since no unloading would be ongoing during this time.

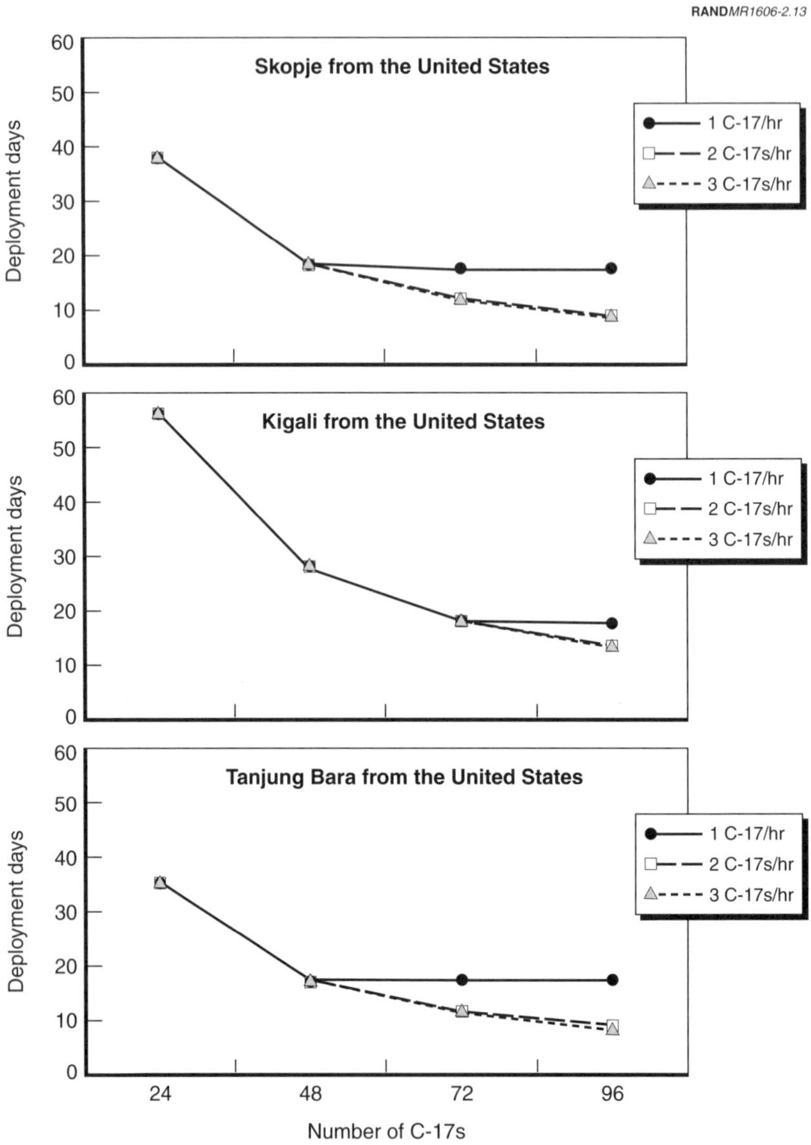

Figure 2.13—Deployment Times by Air from the United States for Each Scenario

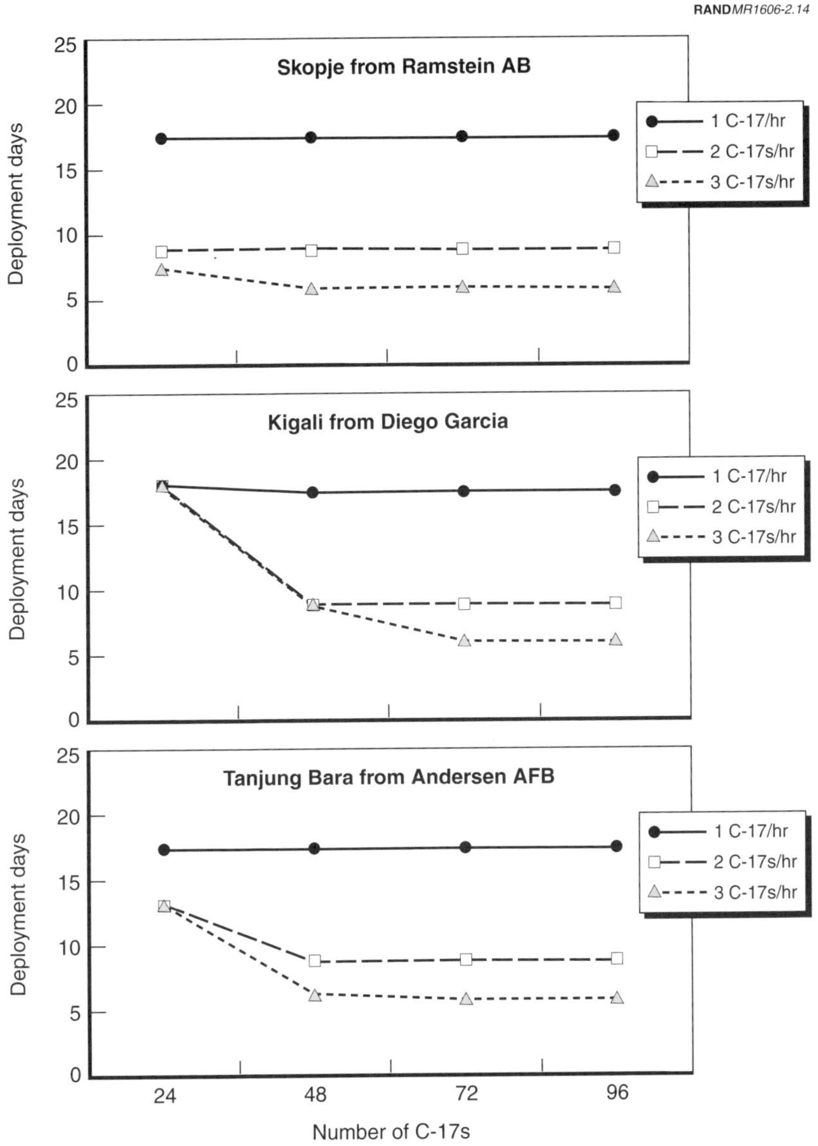

Figure 2.14—Deployment Times by Air from Preposition Sites for Each Scenario

Sealift Deployment Analysis

The SBCT deployment locations and critical distances for deployment by sea for each of the scenarios are summarized in Table 2.7. Note that sealift distances are given in nautical miles and that road-march distances are in statute miles.

Although more ships (or ship sorties) would be required to deploy a unit, the shorter length and shallower draft would allow entry into more ports than are currently available to large MSC ships. This option is represented by the 40-kt shallow-draft case in Figure 2.15, which shows the deployment times from the United States and from preposition sites for various ship types. The base cases are 27-kt FSS from the United States and 20-kt Maritime Prepositioning Ships (MPS) from the preposition (prepo) sites.[50] Production of ships that are capable of 40 kt and similar in range and capacity to the current

Table 2.7

Critical Scenario Information: Sealift

	By Sea					
	From the United States			From Regional Prepo Site		
Deployment Location	SPOD	U.S.–SPOE	Steam Distance (nmi)	Regional-SPOE	Steam Distance (nmi)	Road March (statute miles)
Skopje, Macedonia	Thessaloniki, Greece	Beaumont, TX	6,378	Camp Darby, Italy	1,100	136
Kigali, Rwanda	Mombasa, Kenya	Beaumont, TX	9,200	Diego Garcia	1,900	930
Tanjung Bara, Kalimantan	Tanjung Bara, Kalimantan	Honolulu, HI	5,100	Guam	1,800	0

50As noted earlier in this chapter, all FSS and LMSR ships are based on the Atlantic and Gulf coasts of the United States. This is a DoD policy decision. The ships could be based elsewhere.

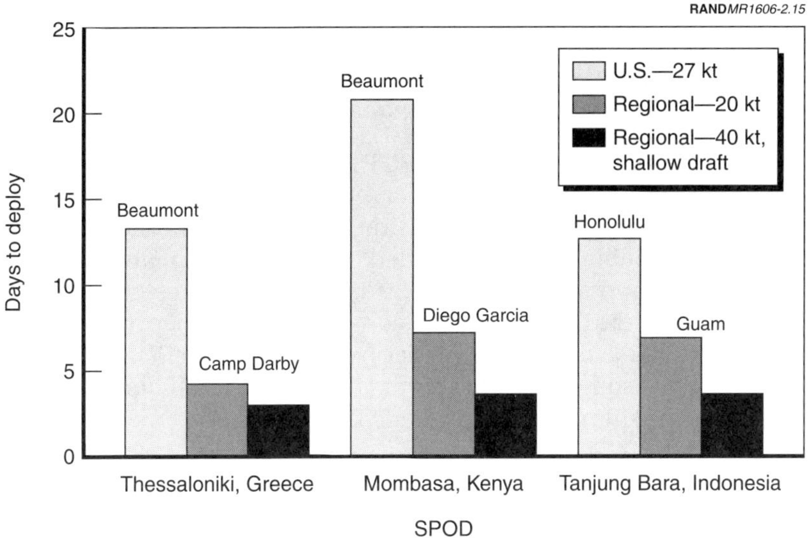

NOTE: The deployment times for 40-kt shallow-draft sealift from regional bases do not include refueling time. Current-generation catamarans can reach about 1,000 nmi at 40 kt with a full load. Our deployment scenarios to Indonesia and Kenya exceed that range. A number of options are available to extend the range of the catamarans, including building longer-range catamarans, trading some payload for additional fuel, making refueling stops at en route ports, and underway refueling. Underway refueling is probably the most robust option, since the capability exists today and does not require permission to use en route ports. If current oilers are used, they would have to put out ahead of the faster catamarans. Alternatively, some catamarans could be converted to oilers, offering the ability to accompany the troop-carrying ships. Based on past experience, underway refueling will likely require the ships to slow to 15–20 kt and could also require a change in heading in rougher seas. Refueling should take 30 minutes to an hour. Best case (20 kt, original heading) would add 15 to 30 minutes to the transit time. In the worst case, the sea state might require the refueling to take place on a heading directly opposite of the desired destination. In that case, the refueling would take place at the slowest speed that still allowed steerage. Assuming 10 kt in the wrong direction for an hour, that would add an hour and 15 minutes to the deployment time. If refueling were required, the catamarans carrying the SBCT would likely depart on a staggered timetable. We thank RAND colleague John Schrader for sharing his expertise and experience in at-sea replenishment and refueling.

Figure 2.15—Port-to-Port Deployment Times (Including Loading and Unloading) for the SBCT

MSC global deployment fleet is years away and may not be technologically or operationally feasible.[51] As discussed in the Shallow-Draft Sealift section of this chapter, current and near-term technology could provide smaller, high-speed vessels with shorter ranges that may be suitable for regional deployments.

The time required to sealift the SBCT in the three scenarios from the United States is shown in Table 2.8 for the 27-kt FSS case. Table 2.9 presents that time from the preposition location for the 40-kt shallow case. An assembly day was included in the time estimates for sealift deployment, and a road march was required for three of the sealift cases because the objective was a considerable distance inland.[52] By comparing these tables, the reader can see the benefits of prepositioning: the steam times are shorter, and no time is required to load the ships (the ships are already loaded).

As discussed earlier in this chapter, we calculated that unloading an SBCT simultaneously from two LMSRs would take 1.7 days. We used this same figure for the shallow-draft prepo ships, although it is likely

Table 2.8

Sealift Deployment Time from the United States (27-kt FSS Case)

Activity	Kosovo (days)	Rwanda (days)	Indonesia (days)
Load ships	1.7	1.7	1.7
Steam	9.8	15.9	7.9
Unload	1.7	3.1	3.1
Assemble	1.0	1.0	1.0
Road march	0.9	7.2	0.0
Total	15.1	28.9	13.7

[51]Even if technological problems can be overcome, development costs—especially if a civilian need does not exist—may make these ships unaffordable.

[52]In the airlift case, the SBCT was brought to the deployment location over the course of the entire time of deployment (piecemeal, in C-17-sized chunks). For sealift, the entire SBCT arrives at the port at the same time. Therefore, we assumed that an additional day would be required on the sealift side to assemble and prepare for the road march.

Table 2.9

Sealift Deployment Time from Preposition Locations (40-kt Shallow-Draft Case)

Activity	Kosovo (days)	Rwanda (days)	Indonesia (days)
Steam	1.3	2.0	1.9
Unload	1.7	1.7	1.7
Assemble	1.0	1.0	1.0
Road march	0.9	7.2	0.0
Total	4.9	11.9	4.6

to be a worse case for the smaller, shallow-draft vessels: In most ports, more than two could unload simultaneously. Also, the smaller vessels can approach and moor (or cast off and clear) in minutes rather than the hours assumed for large ships. The ability to enter small and shallow harbors to unload at the dock would save, on average, 1.5 days that would have been devoted to lighters unloading the larger ships offshore. In foul weather, the smaller ships might be able to dock and unload while the larger ships waited for the calm seas necessary for lighters to operate, offering a potentially huge operational advantage.

In Figure 2.16, which presents graphically the sealift deployment times in Tables 2.8 and 2.9, for ease of comparison, we can see that the deployment times for all prepositioning cases is about half of that from the United States. In addition, we can see that the long road march in Rwanda extends the overall deployment time by almost one week.

Airlift Versus Sealift

The total deployment times for all scenarios are presented in Figure 2.17, with a bar for each of the four deployment cases for comparison: airlift from the United States, sealift from the United States, airlift from the preposition site, and sealift from the preposition site. The shorter bars, which indicate fewer deployment days, are the better options.

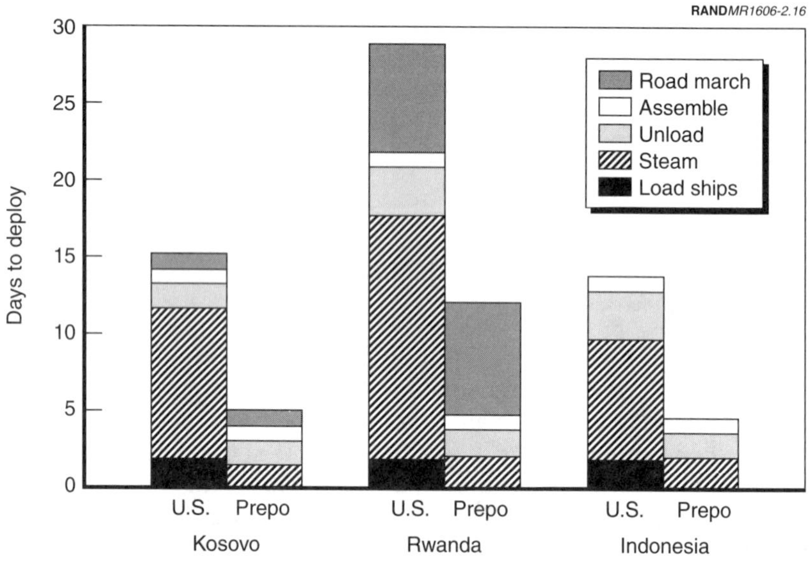

Figure 2.16—Components of Sealift Deployment Time for U.S. and Prepositioning Cases, for Comparison

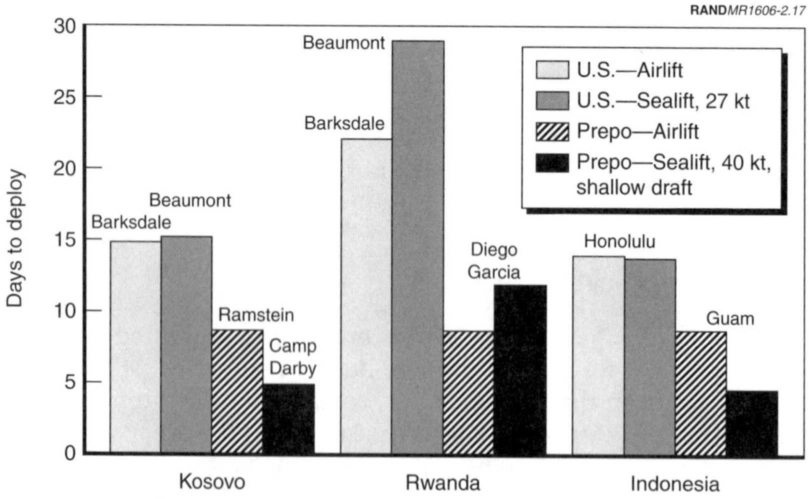

Figure 2.17—SBCT Deployment Times by Air and by Sea from U.S and Preposition Sites

Sixty C-17s were allocated to the operation—over one-quarter of the entire available fleet—and we assumed a throughput of 2 C-17s per hour—a MOG of 3.5 to 4.5, depending on the average time that each C-17 spends on the ground.[53] As discussed earlier in this chapter, a MOG in this range indicates a well-developed airfield. For most SBCT deployments, this level of MOG may not be available: The SBCT is envisioned to be used in contingencies that will likely occur in less-developed, war-torn regions.

In all cases, prepositioning is the single-most effective means for speeding the deployment of the SBCT. Interestingly, in two of the three cases that we studied, the SBCT arrived somewhat faster by sea than by air from the United States. In two of the three cases we studied, the SBCT could deploy fastest from preposition sites by sea; this conclusion is predicated on the use of 40-kt, shallow-draft ships like the catamaran ferries currently being tested by DoD. The combination of faster transit time and the ability to access small ports without the time-consuming use of lighterage is what makes this class of ship so attractive. However, the Rwanda case shows the value of airlift for operations deep in the interior. Indeed, the farther into the interior of the landmass the operation is, the greater is the advantage of airlift over sealift.

---

[53]Two unloading wartime times are shown in the planning factors: 2.25 hours, the regular offload time, and 1.75 hours, the expedite time. See HQ USAF, 1998, Table 5.

Chapter Three

DECISIONS TO INITIATE OPERATIONS

In Chapter Two, we explored the challenge of rapidly moving a medium-weight Army brigade like the SBCT. Our analysis of three representative scenarios suggests that SBCT equipment sets or the units themselves would have to be forward-deployed in at least three regions to achieve response times of under 10 days. In this chapter and the next, we take a step back and try to understand how rapidly an SBCT would need to deploy to be considered a strategically responsive force. We do this by first considering the timelines associated with the political-military decision to initiate joint military operations. In Chapter Four, we then consider whether a global rapid-response capability is essential or whether forces like the SBCT should be postured to respond to crises in specific regions.

This chapter outlines missions that national leaders will expect military forces to perform, assesses how much warning time is likely to be available, explores how the United States typically decides to deploy forces, and identifies factors that influence decisionmaking.

MISSIONS

Ironically, the end of the Cold War has actually increased the frequency with which the United States projects military power. With the collapse of the Soviet Union, the United States suddenly acquired unprecedented military superiority and great freedom of action. At the same time, it found more common ground with China and Russia, especially in the area of peace operations sanctioned by the United Nations Security Council (UNSC). In this new security environment, likely missions for U.S. forces include the following:

- Stopping an aggressor from
  - conquering a country (Korea, 1950; Vietnam, 1964–1973; Kuwait, 1990)
  - dominating a choke point (possible operation: Strait of Hormuz)
- Conducting contingency operations to
  - control a country (Grenada, 1983; Panama, 1989; Haiti, 1994)
  - coerce an opponent (Bosnia, 1995; Kosovo, 1999)
  - protect U.S. citizens (Liberia, 1996)
  - destroy weapons of mass destruction
- Conducting enforcement operations to
  - enforce agreements (Somalia, 1993; Bosnia, 1995 to present; Kosovo, 1999 to present)
  - stop genocide ("might have" operation: Rwanda, 1994)
  - protect humanitarian aid (Somalia, 1992–1993)
- Conducting counterterrorism operations to
  - destroy terrorist groups (al Qaeda, 2001–2002)
  - coerce or overthrow states that support terrorist groups (Libya, 1986; Afghanistan, 2001).

In this section, we describe these missions in terms of force size, warning time, and other factors pertinent to decisionmaking. Force size and warning time are the major conditions for analyzing the "speed" of deployment.

Stopping Aggressors

The first category captures the three largest conflicts fought by the United States since 1945.

Conquering a Country. The Vietnam Conflict involved extensive counterinsurgency operations, but conventional forces from the north posed the greater, and ultimately fatal, threat. Although different in many respects, Korea and Kuwait have some similarities. They were both strategic surprises, although the United States was far better prepared to conduct operations in the Persian Gulf than on the Korean peninsula. In both cases, U.S. forces arrived after friendly territory had been overrun; in both cases, some U.S. forces remained deployed indefinitely to repel a new invasion.

Trying to Dominate a Choke Point. The United States has a strong national interest in enforcing freedom of navigation and freedom to use international airspace worldwide. One example is the Strait of Hormuz, regularly transited by large oil tankers.

Conducting Contingency Operations

The second category captures a variety of missions that U.S. forces accomplish to advance national interests. These missions might be conducted either unilaterally or with assistance from allies.

Control a Country. Examples include Grenada, Panama, and, perhaps, Haiti. In Grenada and Panama, U.S. forces were used to overthrow a government. Haiti might be considered either a contingency operation or an enforcement operation. The United States had a mandate from the UNSC and ultimately entered the country unopposed. However, U.S. forces were fully prepared to enter Haiti against opposition had the Cedras regime remained obdurate. Viewed more broadly, Haiti fits into a pattern of U.S. intervention in the Caribbean and Central America.

Coerce an Opponent. The United States used air power to stop the Bosnian Serbs from bombarding Sarajevo and to make Yugoslavia withdraw its forces from Kosovo.

Protect U.S. Citizens. U.S. forces frequently protect U.S. citizens and evacuate them from areas of conflict. Such evacuation may occur alone or in the context of other operations, as during operations in Grenada.

Destroy Weapons of Mass Destruction. Although it has antecedents in World War II, destruction of weapons of mass destruction (WMD) is a fairly new mission that may become more important in the future. The United States attempted to destroy Iraqi WMD during the Persian Gulf War, but lacked adequate intelligence on Iraqi programs to destroy all weapons, production/research facilities, and launchers. Subsequent attempts to eradicate Iraqi programs through a special UN commission have been unsuccessful.

Conducting Enforcement Operations

The third category captures missions that U.S. forces accomplish under authority of the UNSC and in concert with other countries to enforce international peace. Although humanitarian motives are always present, the United States usually decides to participate on the basis of national interests.

Enforce Agreements. In 1993, the United States supported the UN Operation in Somalia, which attempted unsuccessfully to enforce provisions of the Addis Ababa Accords among the warring factions. In 1995, the United States led a large-scale peace operation to enforce military provisions of the Dayton Agreements, which ended the Bosnian War. In 1999, the United States participated in a large-scale peace operation to enforce an agreement with Yugoslavia regarding Kosovo and to pacify Kosovo pending a political settlement.

Stop Genocide. The United States is a signatory to the Genocide Convention, which states that the contracting parties confirm that genocide is a "crime under international law which they undertake to prevent and punish."[1] However, the United States declined to conduct operations in Rwanda, despite evidence of genocide, leaving the French to conduct Operation Turquoise, which began only after genocide had been perpetrated.

Protect Humanitarian Aid. From December 1992 through March 1993, the United States protected humanitarian aid in Somalia from depredation by warring factions. Thereafter, it withdrew most of its forces, but continued to support a UN operation that had a more ambitious mandate for implementing peace agreements.

---

[1]The UN General Assembly adopted the Convention on the Prevention and Punishment of the Crime of Genocide on December 9, 1948. President Harry S. Truman signed the Convention on December 12, 1948, but Congress did not ratify it until November 25, 1988. The Convention is reflected in U.S. law through the Genocide Implementation Act of 1998, Title 18, Part I, United States Code, also known as the Proxmire Act after Senator William Proxmire (D-Wisconsin).

Conducting Counterterrorism Operations

On September 11, 2001, terrorists seized control of four U.S. airliners and flew three of them into buildings, destroying the World Trade Center in New York City, damaging one section of the Pentagon in Washington, D.C., and causing over 3,000 deaths.[2] The unprecedented scale of these attacks was a shock to the nation, making counterterrorism the highest national priority. In a televised address to the nation on September 11, President Bush said: "We will make no distinction between the terrorists who committed these acts and those who harbor them,"[3] indicating that the United States would take action against states that harbored or harbor terrorists. The United States subsequently conducted such an operation in Afghanistan (Operation Enduring Freedom).

In the past, Special Operations Forces accomplished the active counterterrorism mission; conventional forces protected themselves against terrorism and provided security to civilian activities. As an exception, the United States conducted air attacks on Libya to discourage it from supporting terrorism. In the future, conventional forces are likely to take an increasing role in the active counterterrorism mission.

Destroy Terrorist Groups. In conjunction with Special Operations Forces, conventional units might conduct raids against terrorist groups such as al Qaeda ("The Base"), which is controlled by the Saudi terrorist Osama bin Laden, who was responsible for the September 11 attacks on the United States. Such raids might take days or weeks to complete and include wide-area searches for individual terrorists and their support.

---

[2]As of January 25, 2002, the Federal Emergency Management Agency reported 1,115 bodies recovered and 2,024 registered missing at the World Trade Center in New York City. Another 44 were killed on United Airlines flight 93 when it crashed in Pennsylvania and 189 were killed in the attack on the Pentagon, for a total of 3,372. See http://www.fema.gov/emanagers/nat020202.htm and Eric Lipton, "A New Count of the Dead, but Little Sense of Relief," *The New York Times*, December 2, 2001, available at http://college4.nytimes.com/guests/articles/2001/12/02/887970.xml.

[3]George W. Bush, "Statement by the President in His Address to the Nation," national television, 8:30 p.m. EDT, available at http://www.whitehouse.gov/news/releases/2001/09/20010911-16.html.

Coerce or Overthrow Terrorist States. Also in conjunction with Special Operations Forces, conventional forces might coerce or take down regimes known to harbor terrorists, either directly or through support of their opposition, such as the Northern Alliance, which opposed the Taliban[4] regime in Afghanistan. The United States attempted first to coerce this regime; when coercion failed, U.S. air power was used to support local Afghan forces in their efforts to defeat Taliban forces and evict the Taliban government. The combination of U.S. air power, Special Operations Forces, and Afghani ground forces proved spectacularly successful.

TIMELINE

To analyze decisions to deploy military forces, we employ a simple timeline, or crisis-evolution framework, with the following points: concern, urgent danger, precipitating event, and initiation of military operations. Concern and the perception of urgent danger are waypoints in a process we call "warning." Figure 3.1 depicts this framework, together with characteristic actions that might be taken over time.

This framework is an abstraction from the real world. The real world is less coherent and often messy. Any actual case is likely to present a more complicated chain of events. We discuss these points and aspects underlying these points in the subsections that follow.

WARNING

Decisionmaking depends heavily on *warning*—the judgment that something is happening or about to happen that would prompt the United States to respond militarily. Warning is a difficult concept because it ultimately depends on perceptions. In most instances of strategic surprise, the United States had indications that could have generated a warning. In addition, there can be an enormous gap between strategic warning and operational response. In the most

---

[4]*Taliban* ("students of the Koran") began as a movement of young Afghani, who were scandalized by the civil war that broke out after Soviet occupation and by the lifestyles of competing leaders. Many came from the refugee camps in Pakistan.

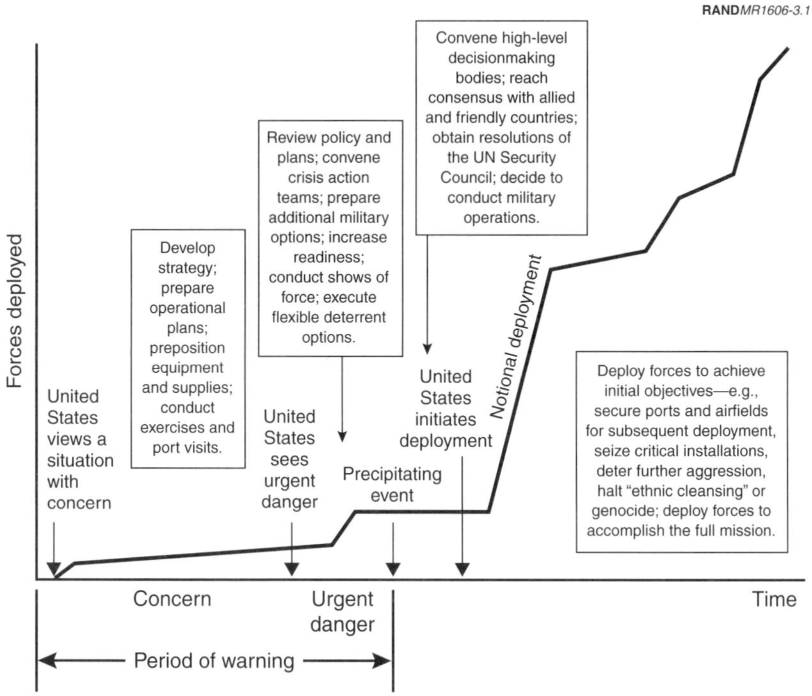

Figure 3.1—Framework for Strategic Deployment

famous example of such a gap, the United States knew in late 1941 that Japan was approaching a decision for war. Indeed, it even knew when Japan would break diplomatic relations. Yet it failed to use this warning operationally. In this instance, U.S. military decisionmakers underestimated Japanese capability and failed to properly assess their own vulnerability. To take another example, Saddam Hussein tried to seize Iranian oil in Khuzestan, but was repelled in a long, bloody war. He threatened Kuwait with invasion, an entirely plausible threat in view of his record—and certainly a clear warning—yet the Gulf States and the United States were still surprised by the invasion. The Gulf States, which were accustomed to propitiating their opponents, were surprised when propitiation failed.

The proposed framework comprises three steps: (1) concern, usually extending over years, (2) urgent danger, usually extending over months, and (3) precipitating event, usually occurring in days. *Concern* implies that if the situation continues to develop unfavorably, it might require military action. *Urgent danger* implies that a situation is so inherently threatening that military action might be required at any time. In some cases—for example, the Iraqi threat to Kuwait, a period of urgent danger could extend over years, even decades.

Concern

Concern begins when U.S. decisionmakers become aware of a situation that could require U.S. military operations eventually. Concern often lasts months or years; it may even last for decades. During this period, the United States might develop and implement a strategy to shape developments, including military engagement, the actions of which might include prepositioning equipment and supplies, developing contingency plans, and conducting military exercises.

U.S. operations in Grenada, Panama, and Haiti are cases in point. During a televised speech delivered in March 1983, President Ronald Reagan denounced construction of an airfield in Grenada with Soviet and Cuban assistance.[5] Nearly seven months later, in October 1983, he ordered execution of Operation Urgent Fury to eliminate the murderous regime of Bernard Coard. Almost two years elapsed between the February 1988 indictment of Noriega by two Florida juries and the December 1989 decision to initiate Operation Just Cause in Panama. Finally, nearly a year elapsed between the October 1993 failure of the U.S.S. *Harlan County* to dock in Port-au-Prince and the initiation of Operation Uphold Democracy in Haiti. In all three cases, there was a substantial time lag between the original events and the decision to initiate military operations.

---

[5]Reagan's March 23rd speech was a broad message about the need for a sizable defense budget to protect America against the Soviet threat. One of his examples was Grenada, about which he stated: "On the small island of Grenada, at the southern end of the Caribbean chain, the Cubans, with Soviet financing and backing, are in the process of building an airfield with a 10,000-foot runway. Grenada doesn't even have an air force. Who is it intended for? . . . The Soviet-Cuban militarization of Grenada, in short, can only be seen as power projection in the region." See http://www.reagan.utexas.edu/resource/speeches/1983/32383d.htm. Last accessed on November 20, 2002.

Urgent Danger

Urgent danger begins when U.S. decisionmakers become aware of a crisis or conflict that is likely to require U.S. military operations quite soon. During this period, the United States might review its policy and take actions to prevent or allay the threatening conflict. Military actions might include increased readiness and shows of force, such as flexible deterrent options.

The concept of *urgent danger* is useful, but also problematic. If urgent danger is protracted, it can be like the little boy who cried wolf; the false alarms tend to inure the hearer, and urgency is lost. For example, U.S. observers failed to anticipate the invasion of South Korea in 1950, in part because very high levels of tension, involving cross-border operations and guerrilla warfare in the South, had not led to war in the past. When a country implodes in civil war, it may give short-term warning for months or even years before U.S. decisionmakers feel impelled to act, as in Bosnia and Kosovo. In some sub-Saharan countries, such as the Congo, Burundi, Liberia, Somalia, and the Sudan, conflict has continued so long that the distinction between *concern* and *urgent danger* may have been obliterated.

Precipitating Events

Precipitating events spur U.S. decisionmakers to initiate military operations.

What spurs a U.S. decisionmaker to initiate military operations is the volition of the participants. At one extreme, the enemy's decision initiates military operations, leaving the United States to respond—for example, Korea in 1950 and Kuwait in 1990. At the other extreme, the United States decides when military operations will begin, leaving the opponent to respond. In Haiti, President Clinton's 1994 ultimatum to the military junta was the precipitating event. In other cases, both sides have engaged in a pattern of challenge and response, leading to U.S. military operations: Vietnam (1962–1964), Panama (1988), Bosnia (1992–1995), and Kosovo (1998–early 1999). In these cases, volition ultimately lay with the United States, which made deliberate decisions to intervene.

Most precipitating events are unambiguous, but some may be controversial. For example, what event precipitated deeper U.S. involvement in Bosnia? The fall of Srebrenica was such an event, because it prompted NATO to take a stronger stance, leading directly to Deliberate Force, a U.S.–led air effort against the Bosnian Serbs. But another candidate would be Operation Storm, the Croatian assault on Krajina, which compelled the Bosnian Serbs to sue for peace.

Timelines for Selected Operations

In Table 3.1, we put dates on the three phases of crisis evolution discussed above, for selected military operations. This is inherently a subjective exercise.

Concern for Afghanistan might be dated to earlier incidents, such as the attacks on the World Trade Center on 26 February 1993 and the Khobar Tower in Dhahran on 26 June 1996. However, it certainly began no later than 7 August 1998, when al Qaeda attacked U.S. Embassies in East Africa. In response to those bombings, U.S. forces struck al Qaeda training facilities in Afghanistan and a factory in the Sudan thought to be engaged in producing chemical weapons. Urgent danger certainly began no later than President Clinton's finding:

> I, WILLIAM J. CLINTON, President of the United States of America, find that the actions and policies of the Taliban in Afghanistan, in allowing territory under its control in Afghanistan to be used as a safe haven and base of operations for Usama bin Laden and the al Qaeda organization who have committed and continue to commit acts of violence against the United States and its nationals, constitute an unusual and extraordinary threat to the national security and foreign policy of the United States and hereby declare a national emergency to deal with that threat.[6]

In light of this record, the United States usually is in the concern phase for at least one year, but more typically two years or more. It usually spends months to years in the urgent-danger phase, but

---

[6]William J. Clinton, *Executive Order 13129 of July 4, 1999, Blocking Property and Prohibiting Transactions with the Taliban*, Washington, D.C.: Presidential Documents, *Federal Registry*, Vol. 64, No. 129, 7 July 1999.

Table 3.1

Crisis Evolution for Selected Operations

Country	Concern Began	Urgent Danger Began	Precipitating Event Occurred	Military Operation Began
Korea	1948 Democratic People's Republic of Korea (DPRK) is established under Kim Il Sung.	Early Jun 1950 U.S. intelligence estimates that DPRK is prepared to invade at any time.	25 Jun 1950 DPRK invades the Republic of Korea.	27 Jun 1950 President Truman orders air and naval support to Republic of Korea forces; 30 Jun 1950 Truman authorizes deployment of U.S. land forces to Korea.
Vietnam	1954 Geneva Conference divides Vietnam at the 17th parallel.	1962 U.S. begins advising the South Vietnamese military.	4 Aug 1964 Gulf of Tonkin incident	8 Mar 1965 9th Marine Expeditionary Brigade lands at Da Nang.
Grenada	1979 Leftist New Jewel movement seizes power in Grenada.	1982 Grenada begins expansion of Salines Airport with Cuban assistance.	19 Oct 1983 Marxist faction murders Prime Minister Maurice Bishop and five associates.	25 Oct 1983 D-Day Operation Urgent Fury
Libya	1985 Muammar Qaddafi praises Abu Nidal's terrorism.	24 Mar 1986 Libya fires SA-5 missiles at U.S. F-14s.	5 Apr 1986 Bomb explodes in a Berlin discotheque, injuring U.S. soldiers.	14 Apr 1986 D-Day Operation El Dorado Canyon
Panama	1986 Violent demonstrations against regime of Manuel Noriega; U.S. Senate passes resolution urging democracy.	Feb 1988 Two federal grand juries indict Noriega for drug smuggling; Oct 1989 Noriega crushes a military revolt.	16 Dec 1989 Panamanian forces kill U.S. Lt. Robert Paz.	20 Dec 1989 D-Day Operation Just Cause

Table 3.1—continued

Country	Concern Began	Urgent Danger Began	Precipitating Event Occurred	Military Operation Began
Kuwait (1)	1961 Kuwait asks for British protection against Iraq.	16 Jul 1990 Iraq delivers very strong note to Kuwait, demanding cancellation of its war debts.	1 Aug 1990 Iraq invades Kuwait.	7 Aug 1990 C-Day Operation Desert Shield
Kuwait (2)	7 Aug 1990 C-Day Operation Desert Shield.	31 Oct 1990 NCA approves deployment of VII Corps to theater.	12 Jan 1991 U.S. Congress passes resolution authorizing use of military force.	16 Jan 1991 D-Day Operation Desert Storm
Somalia	1991 Barre regime overthrown; clan-based factions compete for power.	Apr 1992 UNSC establishes first UN Operation in Somalia (UNOSOM I); Aug 1992 U.S. begins Operation Provide Relief.	25 Oct 1992 U.S. suspends Operation Provide Relief when an aircraft is struck by fire.	5 Dec 1992 D-Day Operation Restore Hope
Haiti	1991 Army overthrows President Aristide; U.S. freezes regime's assets.	Oct 1993 USS *Harlan County* departs Port-au-Prince without off-loading; UNSC imposes sanctions on Haiti.	31 Jul 1994 UNSC authorizes use of force to depose the military regime; 15 Sep President Clinton gives ultimatum.	19 Sep 1994 D-Day Operation Uphold Democracy
Rwanda	1962 Belgian protectorate ends; fighting breaks out between Hutus and Tutsis.	Oct 1990 Rwandan Patriotic Front invades northern Rwanda from Uganda.	7 Apr 1994 Killings begin in Kigali; extremist Hutus kill 10 Belgian soldiers trying to protect moderate Hutus.	22 Jun 1994 France begins Operation Turquoise; 17 Jul 1994 U.S. begins Operation Support Hope.

Table 3.1—continued

Country	Concern Began	Urgent Danger Began	Precipitating Event Occurred	Military Operation Began
Bosnia and Herzegovina	1991 Croatia and Slovenia declare independence; war breaks out between Croatia and Serbia.	Mar 1992 Bosnia votes for secession; fighting breaks out in Sarajevo.	6 Jul 1995 Bosnian Serb forces attack the "safe area" of Srebrenica.	20 Dec 1995 Operation Joint Endeavor formally assumes responsibility for military operations in Bosnia and Herzegovina.
Liberia	1990 Rebels topple the Samuel Doe regime.	1991 Liberia begins descent into clan warfare.	1 Apr 1996 People seek refuge in U.S. Embassy and housing area.	9 Apr 1996 D-Day Operation Assured Response
Kosovo	1989 Serbia rescinds autonomy of Kosovo Province.	Feb 1998 Serbian offensive in Kosovo slaughters civilians.	15 Jan 1999 Serb forces massacre civilians at Racak; U.S. diplomat William Walker visits the site.	24 Mar 1999 D-Day Operation Allied Force
Afghanistan	7 Aug 1998 Bombs explode in U.S. Embassies in Nairobi and Dar es Salaam; al Qaeda is implicated.	4 Jul 1999 In Executive Order 13129, President Clinton finds that Taliban offers safe haven to al Qaeda and imposes sanctions.	11 Sep 2001 Al Qaeda crashes airliners into World Trade Center and Pentagon, killing 3,000 people.	7 Oct 2001 D-Day Operation Enduring Freedom

sometimes only weeks. In some cases, there may be little long-term concern prior to a crisis, but years in the urgent-danger phase, reflecting a state of continual crisis and conflict. Bosnia and Liberia are cases in point.

Despite spending at least a year in the concern phase and at least a few weeks in the urgent-danger phase, the United States was highly surprised in four cases, including two that involved theater-level conflict: Korea and Kuwait. Moreover, both these conflicts demanded rapid deployment of large forces over intercontinental distances to prevent a small ally from being overrun, which indeed happened in both instances. Defense of a small ally sets the most demanding standard for projection of U.S. forces.

Table 3.2 summarizes the time U.S. planners and decisionmakers have spent in the concern and urgent-danger phases in selected military operations since 1945.

THE CRISIS DECISIONMAKING PROCESS

Military forces have seldom begun to deploy precisely when conflict appeared imminent or precisely when a precipitating event occurred, owing in part to the decisionmaking process. Political decisionmaking can occur in hours—for example, President Truman's decision to resist the Communist invasion of the Republic of Korea (1950) or President Reagan's decision to intervene in Grenada (1983). But it usually takes longer and may extend over months—for example, President Clinton's decision to restore the Aristide government to Haiti (1994). In this case, the Clinton administration was anxious to avoid a forcible intervention, preferring to negotiate with the military junta led by Raoul Cedras. Political-military decisionmaking often consumes a significant amount of time: to explore alternatives to combat operations and to weigh the threat to U.S. interests, especially considering the risk of U.S. casualties. These time needs are discussed below, in turn.

Table 3.2

Concern and Urgent-Danger Phases for Selected Military Operations

Country	Concern	Urgent Danger
Korea	2 years	3 weeks
Vietnam	8 years	18 months
Grenada	3 years	1 year
Libya	1 year	2 weeks
Panama	2 years	2 months
Kuwait	3 decades	2 weeks
Somalia	1 year	1 month
Haiti	2 years	10 months
Rwanda	2 decades	5 months
Bosnia	1 year	3 years—protracted war
Liberia	1 year	5 years—anarchic conditions
Kosovo	9 years	10 months
Afghanistan	2 years	14 months

Exploring Alternatives

American decisionmakers often seek alternatives to large military operations, including diplomacy, threat of military force, and covert operations. As Caspar Weinberger argued: "The commitment of U.S. forces to combat should be a last resort."[7] Indeed, the decision to use force has generally been considered a last resort, only after lesser measures have failed. Thomas Schelling notes: "It is the *threat* of damage, or of more damage to come, that can make someone yield or comply. It is *latent* violence that can influence someone's choice—violence that can still be withheld or inflicted, or that the victim believes can be withheld or inflicted."[8]

During the Cuban missile crisis, for example, the deployment of a naval blockade around Cuba—in addition to such military actions as ordering the B-52 bomber force into the air, placing missile crews on maximum alert, and moving U.S. troops into Florida and the southeastern United States—helped *coerce* the Soviet Union into removing its missiles without the United States resorting to combat operations. Moreover, the Nixon administration's decision to increase the readiness of U.S. forces to Defense Condition III—place the 82nd Airborne Division on alert, recall B-52 aircraft based in Guam to the United States, and move aircraft carriers to the Mediterranean—helped deter the Soviet Union from deploying troops in the Middle East during the 1973 Yom Kippur War.[9] In other cases, the United States used covert action to coerce or deter opponents, as in Guatemala in 1954, the Bay of Pigs in 1961, and Chile in 1973.

Threats to Important U.S. Interests

Threats to important and vital U.S. interests will usually imply more-rapid decisionmaking than in exploring alternatives. Some actions

---

[7]Caspar Weinberger, *Fighting for Peace: Seven Critical Years in the Pentagon,* New York: Warner Books, 1991, p. 454.

[8]Thomas C. Schelling, *Arms and Influence,* New Haven, Conn.: Yale University Press, 1966, p. 3. See also Barry M. Blechman and Stephen S. Kaplan, *Force Without War: U.S. Armed Forces as a Political Instrument,* Washington, D.C.: The Brookings Institution, 1978.

[9]See Henry Kissinger, *Years of Upheaval,* Boston, Mass.: Little, Brown and Company, 1982, pp. 575–591.

threaten U.S. interests directly. President Harry Truman quickly decided to repel North Korean aggression because he was determined to stop the spread of Communism, which threatened to engulf Europe and Asia. President George Bush quickly decided to defend Saudi Arabia because Saddam Hussein's aggression would almost certainly have continued if left unchecked, and unimpeded access to Gulf oil was vital to the world's prosperity.

Other actions threaten U.S. interests indirectly. For example, Serb oppression of Albanians in Kosovo threatened to upset NATO's position in the Balkans and, therefore, destroy the credibility of an alliance that has long been the foundation of U.S. security policy in Europe. More immediately, NATO countries were disturbed by sudden influxes of Albanian refugees.

In contrast, the United States has often refrained from using military force where it had no important interests at stake, even in the face of genocide. Two examples illustrate this point. There was little support in Washington—particularly on Capitol Hill—for stopping the genocide in Cambodia between 1975 and 1979, when the Khmer Rouge murdered at least 1.7 million people.[10] Nor did the United States try to stop the murder of some 800,000 Tutsi and moderate Hutus during a genocide conducted in Rwanda in early 1994. As Republican Senate minority leader Bob Dole remarked: "I don't think we have any national interest there. The Americans are out, and as far as I'm concerned, in Rwanda, that ought to be the end of it."[11]

Risks of U.S. Casualties

Decisionmakers may hesitate to initiate operations that could entail substantial U.S. casualties. Public-opinion research strongly suggests that expectations of U.S. casualties affect the views of the public, the Congress, and, especially, the executive decisionmakers. In a

---

[10]These figures come from the Cambodian Genocide Program at Yale University.

[11]Samantha Power, "Bystanders to Genocide: Why the United States Let the Rwandan Tragedy Happen," *The Atlantic Monthly*, Vol. 288, No. 2, September 2001, pp. 84–108. See also Philip Gourevitch, *We Wish to Inform You That Tomorrow We Will Be Killed with Our Families: Stories from Rwanda*, New York: Farrar, Straus and Giroux, 1998.

number of U.S. military operations, including those in Somalia, Lebanon, the Dominican Republic, Vietnam, and Iraq, higher casualties, whether estimated or actual, tended to reduce public support.[12] However, willingness to accept casualties seems strongly correlated with U.S. interests. When vital or important interests are at stake, all are more willing to accept casualties. For example, during the Civil War and both world wars, public support remained strong despite casualties. Conversely, when no important interests are at stake, all are less likely to believe that substantial casualties are justified. Moreover, leadership plays an important role:

> The simplest explanation consistent with the data is that support for U.S. military operations and the willingness to tolerate casualties are based upon a sensible weighing of benefits and costs that is heavily influenced by consensus (or its absence) among political leaders.[13]

Decisionmakers are especially concerned about casualty *estimates*. During Operation Allied Force, for example, there was intense debate within the Clinton administration about U.S. casualties that might be incurred by ground forces. General Wesley K. Clark, then Commander in Chief, U.S. European Command, recalled that Secretary of Defense William Cohen warned him before an April 1999 NATO meeting: "Nothing about ground forces. We have to make this air campaign work, or we'll both be writing our resumes."[14] Concerns about casualties and how casualties will affect public support also underlay the decision not to employ Task Force Hawk.[15] To take another example, when the Bush administration debated whether to execute Operation Just Cause, National Security Adviser

---

[12]See Eric V. Larson, *Casualties and Consensus: The Historical Role of Casualties in Domestic Support of U.S. Military Operations,* Santa Monica, Calif.: RAND, MR-726-RC, 1996; John E. Mueller, *War, Presidents and Public Opinion,* New York: John Wiley & Sons, 1973; John E. Mueller, *Policy and Opinion in the Gulf War,* Chicago, Ill.: University of Chicago Press, 1994; James Burk, "Public Support for Peacekeeping in Lebanon and Somalia: Assessing the Casualties Hypothesis," *Political Science Quarterly,* Vol. 114, No. 1, 1999, pp. 53–78.

[13]Larson, 1996, p. xv.

[14]General Wesley K. Clark, 2001, p. 269.

[15]"Having gone to great lengths to deploy TF Hawk, why did the United States decline to employ it? Ultimately, it was because decision-makers perceived the risks to outweigh the potential benefits." Nardulli et al., 2002, p. 94.

Brent Scowcroft insisted on discussing casualty estimates before reaching a decision.[16]

HISTORICAL RESPONSE TIMES

Response time runs from a precipitating event to the moment when the United States must start military operations to prevent undesirable outcomes, such as the occupation of a friendly country, massacre of innocent civilians, or additional terrorist attacks on the United States. Political considerations, especially the need to hold a coalition together, may also add urgency.

Depending on the situation, the United States might prepare for military operations before the precipitating event occurs, including forward positioning of military units. The most important factors affecting preparation are surprise and initiative. *Surprise* is the degree to which a precipitating event catches decisionmakers unawares. *Initiative* implies having freedom of action, especially freedom to choose the time and place at which military operations will occur.

The worst case would be high surprise, when the opponent has the initiative and the United States must respond, as in Korea and Kuwait. The best case would be low or no surprise when the United States has the initiative, as in Libya, Panama, and Haiti. A neutral case would imply no surprise and initiative on both sides of the conflict, as during the period of U.S. involvement in Vietnam.

A very uncertain relationship exists between warning and strategic surprise. Perversely, a very long period of warning can actually contribute to surprise: When years or even decades pass peacefully, it becomes harder to imagine that war will ever come. Kuwait is a case in point. Iraq threatened this small principality from its very creation in 1961, claiming that Kuwait should rightfully have been an Iraqi province. But when invasion finally occurred in 1990, it surprised the Saudis and Kuwaiti governments, which believed that Saddam Hussein was bluffing. General H. Norman Schwarzkopf, the

---

[16]Colin Powell, with Joseph E. Persico, *My American Journey: An Autobiography*, New York: Random House, 1995, p. 425; Bob Woodward, *The Commanders*, New York: Simon & Schuster, 1991, pp. 169–170.

Commander in Chief of U.S. Central Command, thought Iraq would invade, but seize only the Rumaila oil field and Bubiyan Island.[17] Schwarzkopf's assessment erred in ascribing to Iraq a more modest strategy than it actually pursued. In the end, Saddam Hussein achieved surprise largely because the Gulf States and their protector underestimated both his ruthlessness and his poor judgment.

Some missions are urgent because the cost of delay would be high. Examples include an attack on the U.S. homeland or its civilians abroad, aggression against an allied or friendly country, some humanitarian catastrophes, and defense against weapons of mass destruction. Attack on the U.S. homeland would demand an immediate response to prevent further damage and to manage the consequences. For example, part of the immediate response to the terrorist attacks on September 11, 2001, was to fly combat air patrols over likely target areas. A hunt for Osama bin Laden and other members of the al Qaeda terrorist network was also urgent, both to maintain momentum among coalition members and to prevent further attacks. The United States has an open-ended commitment to protect its nationals traveling or residing in foreign countries, usually through their timely evacuation. Since the United States is reluctant to disrupt its relations with foreign countries, noncombatant evacuation may be postponed until the last moment, then conducted on an urgent basis.

Aggression against an allied or friendly country may require an urgent response to prevent further aggression and to exploit the window of opportunity for eliciting support. Following the Iraqi invasion of Kuwait in August 1990, General Schwarzkopf advised King Fahd of Saudi Arabia of the imminent danger to his kingdom:

> I explained that while we didn't know whether the Iraqis intended to attack Saudi Arabia, we judged from their deployment and from similar Iraqi actions during the Iran-Iraq war that they were in what we called a strategic pause, busy rearming and reequipping before

---

[17]H. Norman Schwarzkopf, with Peter Petre, *General H. Norman Schwarzkopf: The Autobiography: It Doesn't Take a Hero,* New York: Bantam Books, 1993.

> continuing offensive operations. They had their best units forward, posed to attack; their posture was certainly not defensive.[18]

Humanitarian catastrophes, especially those caused by "ethnic cleansing" and genocide, usually demand urgent responses to prevent additional loss of life. Some effects, such as forced relocations, are reversible; others, such as the destruction of property and loss of life, are not. For example, the United States and its NATO allies secured the return of hundreds of thousands of Kosovar Albanians to their homes following the withdrawal of Yugoslav troops, but did not prevent the murder and rape of many Kosovars or the looting and razing of their homes and villages. Even if the effect is reversible, it may have serious implications for U.S. policy. For example, "ethnic cleansing" in Kosovo created so much animosity between Serbs and Albanians that the U.S. goal of a well-governed, multi-ethnic Kosovo became unachievable over the short term. Other effects are permanent, such as the genocide that occurred in Rwanda in April 1994 with unanticipated speed and ferocity.

Defense against weapons of mass destruction is a vital interest that may require a very quick response. An adversary may build or acquire weapons of mass destruction to use against the United States. A dramatic example was the introduction of medium-range ballistic missiles to Cuba in 1962. President Kennedy stated that an immediate response was critical because the "urgent transformation of Cuba into an important strategic base—by the presence of these large, long-range, and clearly offensive weapons of sudden mass destruction—constitutes an explicit threat to the peace and security of all the Americas."[19] However, at the height of the Cold War, neither the Soviet Union nor the United States was able to defend itself against the other's nuclear forces and, therefore, followed policies of mutual assured destruction. Since the end of the Cold War, the risk of nuclear conflict between Russia and the United States has receded almost to zero, whereas the risk of attack from smaller adversaries, especially terrorist groups, has increased.

---

[18]Schwarzkopf, 1993, p. 353.

[19]John F. Kennedy, Address to the Nation, broadcast on October 22, 1962.

For selected military operations, Table 3.3 gives rough assessments of surprise, initiative, undesirable outcome, and time to respond measured from the precipitating event.

In five cases (Grenada, Libya, Panama, Somalia, Haiti), the United States had the initiative and chose the time to initiate military operations. Moreover, in all of these cases except Somalia, undesirable outcomes were either remote or not time-sensitive. For example, the United States feared that the regime of Bernard Coard in Grenada might take U.S. citizens hostage—a threat that never materialized. In two cases in which both sides had scope for initiative (Bosnia and Kosovo), the United States still had ample time to prepare for military operations. For Kosovo, the United States and its NATO allies chose to conduct an air effort. Had they chosen to initiate opposed

Table 3.3

Surprise, Initiative, Outcome, Time to Respond for Selected Military Operations

Country	Surprise	Initiative	Undesirable Outcome	Time to Respond
Korea	high	opponent	Communist Korea; threat to Japan	2–3 weeks
Vietnam	none	both	Communist Vietnam; threat to Laos, Cambodia	months
Grenada	high	U.S.	Coard regime	days
Libya	low	U.S.	More terrorist attacks	days
Panama	low	U.S.	Noriega regime	weeks
Kuwait	high	opponent	Iraqi invasion of Saudi Arabia	days
Somalia	none	U.S.	Morbidity due to conflict	weeks
Haiti	none	U.S.	Cedras regime	days
Rwanda	high	opponent	Genocide, morbidity	1–2 weeks
Bosnia	low	both	Violations of Dayton Agreements, resurgence of conflict	weeks
Liberia	low	opponent	Mistreatment of U.S. nationals	days
Kosovo	low	both	“Ethnic cleansing,” massacre of Albanians	months
Afghanistan	low	both	More terrorist attacks	weeks

land operations, they would have had months to prepare, assuming that the operation would have to begin before winter.

In cases involving terrorism (Libya, Afghanistan), preventing more terrorist attacks was an obvious, first-order motive for the United States to mount military operations quickly. But there were also other motives for speed: for Libya, to establish a connection between the Reagan administration's action and the bombing of a Berlin discotheque frequented by U.S. personnel; in Afghanistan, to consider the dynamics of coalition building, the military situation of opposition groups in Afghanistan, the dimensions of human suffering in the region, and the impending onset of winter weather.

This brief survey suggests that the United States is seldom surprised at the strategic level, although the precipitating event may be unexpected and shocking, as were the attacks on September 11. Even when such surprises occur, the United States can often set its own pace and deliberately build up forces. It would be unrealistic to depict the United States as being often surprised by events that demand very rapid deployments. On the contrary, U.S. military preeminence allows its decisionmakers considerable latitude, and they seldom have to respond pell-mell. However, having such latitude does not imply that rapid deployment is not valuable. Often prevailing are strong moral and political motives to proceed as rapidly as possible, short of incurring excessive risk. Quick resolutions tend to reduce the suffering of innocent humans and to minimize political complications, especially within coalitions encompassing disparate views. Finally, note that this analysis based on historical precedents excludes consideration of new capabilities. For example, the United States might have taken a different course of action in Afghanistan had faster-deploying Army forces been available.

Chapter Four

REGIONS OF INTEREST

The Air Force has global reach, operating through air and space. One of its core competencies is *rapid global mobility,* defined as "the ability to rapidly position forces anywhere in the world."[1] The Navy and Marine Corps define their capabilities in reference to the world's oceans and littorals, particularly the capability to "project precise power from the sea."[2] The Army intends to develop an Objective Force with "deployability that will enable us to place a combat-capable brigade anywhere in the world in 96 hours; put a division on the ground in 120 hours; and five divisions on the ground in 30 days."[3] The Army envisions conducting land operations globally: "The Army, supported by the Air Force, has a forcible entry capability that allows it to conduct land operations anywhere in the world."[4] "Strategic responsiveness requires Army forces trained, organized, and equipped for global operations, and commanders and units proficient at force projection."[5]

---

[1]The Honorable F. Whitten Peters and General Michael E. Ryan, *America's Air Force Vision 2020: Global Vigilance, Reach, and Power,* Washington, D.C.: U.S. Air Force, 2001, available at www.af.mil/vision, not paginated.

[2]Admiral Frank B. Kelso II and General C. E. Mundy, Jr., *. . . From the Sea: Preparing the Naval Service for the 21st Century,* Washington, D.C.: U.S. Navy and U.S. Marine Corps, 1992, not paginated.

[3]Headquarters, Department of the Army, *The Army Vision: Soldiers on Point for the Nation, Persuasive in Peace, Invincible in War,* Washington, D.C., August 30, 2001b, available at www.army.mil/vision.

[4]Headquarters, Department of the Army, *The Army,* Washington, D.C.: Field Manual 1, June 14, 2001a, p. 4.

[5] Headquarters, Department of the Army, 2001d, p. 3-19.

The Air Force can rapidly airlift land forces anywhere in the world if those forces are sufficiently light or sufficiently small. In 1983, the Air Force lifted two brigades of the 82nd Airborne Division to Grenada during Operation Urgent Fury. In 1994, the United States could have airlifted a brigade of light infantry to Rwanda to stop the killings.[6] In 1993, the Air Force lifted a heavy battalion task force to Somalia in the aftermath of the October 3 firefight that killed 18 Americans. In 1999, the Air Force lifted Task Force Hawk, a relatively modest force, a short distance to Albania (Ramstein AB in Germany to Rinas Airport in Albania) during Allied Force. But to do so required 442 C-17 sorties and took about one month.[7]

Requirements to deploy large land forces are ultimately a function of U.S. interests and strategy. In the event of war against the Warsaw Pact, NATO expected that the Central Front would be the crucial area of operations. The Central Front extended from the Baltic Sea along the inter-German border (the border between the Federal Republic of Germany and the Democratic Republic of Germany) and the eastern border of Austria to the Alps. During the Cold War, the Central Front dominated mobility requirements, leading Army forces to preposition unit sets of equipment in Europe and to create a Reserve Fleet that included *Algol* Class (SL-7) fast vehicle cargo ships based on the Atlantic seaboard. Since the end of the Cold War, these requirements have become more diffuse, but they are not undifferentiated (i.e., they are not equally applicable in all regions and at all times). U.S. forces are not as likely to operate in South America or Central Asia as they are in Southwest Asia or the Pacific Rim. We recognize that it is impossible to precisely forecast in what countries U.S. forces may operate in the future. Even when an event is widely expected, as was the violent collapse of Yugoslavia, the U.S. response

---

[6]See Scott R. Feil, *Preventing Genocide: How the Early Use of Force Might Have Succeeded in Rwanda*, New York: Carnegie Commission, 1998. Lt. Gen. (Canada) Romeo A. Dallaire, the commander of UN forces in Rwanda when the genocide began, believed that 5,000 modern light infantry arriving in mid-April could have significantly altered the outcome. However, only the United States could have deployed such a force, and the United States was not disposed to intervene. See Power, 2001, pp. 84–108.

[7]Task Force Hawk totaled 7,745 passengers and 22,937 short tons of cargo. At any given time, approximately 12 C-17s and 28 aircrews were involved. Competing requirements limited C-17s to only two parking slots at Rinas Airport. See Headquarters, U.S. Air Force Europe, 1999.

to that event may be highly uncertain. But at a higher level of aggregation, especially at the regional level, patterns become apparent.

The United States is likely to accomplish joint operations in certain regions of interest in the future. In this chapter, we identify such regions on the basis of historical operations involving U.S. forces and of other indications, such as current crises and conflicts, current UN operations, and states sponsoring terrorism. For each region, we present a brief analysis of past operations, accessibility, and security issues that might trigger future operations.

The United States is the only country capable of airlifting large joint forces.[8] This capability can be a trump card, as during operations in Afghanistan, when access was politically constrained and the area of operations was landlocked. Only airlift offers truly global mobility, implying the ability to reach deep into continental landmasses where good land lines of communication are lacking. However, airlift is likely to remain scarce and expensive. It therefore pays to consider where regional patterns will allow other approaches, such as sealift, forward stationing, and prepositioning of equipment and supplies. That said, we first look at how patterns of interest are discerned.

DISCERNING PATTERNS OF INTEREST

Forecasting where U.S. forces are most likely to operate in the future can be done in at least three ways. First, the historical pattern of U.S. operations is the important indicator of likely areas of operations. Second, areas of current crises and conflicts suggest where instability might affect U.S. interests. Third, terrorist groups and the states that sponsor terrorism might be targets for U.S. military operations. We examine each way in the following subsections.

---

[8]A *large joint force* is defined for study purposes as at least one ground force brigade equivalent and one air wing equivalent. Such a force might include a Stryker Brigade Combat Team, a Marine Expeditionary Unit, an Air Force Air Expeditionary Wing, and Navy forces such as a Carrier Battle Group and an Amphibious Ready Group.

U.S. Operations Since 1945

Almost 60 years have passed since 1945 and over a decade since the end of the Cold War, certainly enough time to establish a pattern that indicates where U.S. forces are most likely to operate. Figure 4.1 displays significant U.S. military operations since 1945.

These operations are clustered in certain regions: the Caribbean and Central America, Western Europe, North Africa, East-Central Africa, the Middle East, Southwest Asia, and the Pacific Rim. Obviously, the relative number of operations does not reflect the level of interest, if only because two conflicts on the Pacific Rim (Korea and Vietnam) dwarf all other operations in cost, duration, and U.S. casualties. No significant operations took place in South America, other parts of Africa, other parts of Asia, and the Australian continent.

The United States does not respond to crises and conflicts indiscriminately. Rather, it responds in accordance with its interests and the interests of its friends and allies: for example, choosing to enforce peace in Kosovo because NATO was directly concerned, but not in Rwanda, although far more people were at risk in that country.

RAND*MR1606-4.1*

Figure 4.1—Selected U.S. Operations Since 1945

Comparing areas of past operations and commitments to areas of current crises/conflicts offers some insight into which of those latter areas the United States is most likely to intervene. (See Figure 4.2 in the next subsection.)

Areas of Current Crises and Conflicts

The United States might also apply military force in areas where its forces have not operated previously. It might even choose to employ military forces where it has no strong national interests. Indeed, the United States had no interest in Somalia in 1992–1993, other than to prevent starvation, the original motive for intervention. The United States might, for example, deploy forces under a UNSC resolution to prevent or allay conflict in distant countries, simply to promote international peace and security, the original purpose of that organization.

A quick review of current crises and conflicts reveals overlap with areas of historical U.S. military involvement, but also wide divergence. On the one hand, Korea, China-Taiwan, and the Near East are overlaps. On the other hand, the United States has thus far avoided military involvement in some of the world's most violent and dangerous conflicts. Although active diplomatically, the United States has not deployed forces to Africa south of the Sahara, with the exception of the debacle in Somalia. Nor have large U.S. forces been deployed to South America, although the United States is clearly interested in the outcome of Colombia's internal conflict. Figure 4.2 depicts areas of current crises and conflicts.

The War on Terrorism

In response to the terrorist attack on September 11, 2001, the United States announced a war against global terrorism and against states that sponsor global terrorism. Less than a month after the attack, the United States initiated military operations in Afghanistan, a venue that was highly unlikely before the September attack. In its war on global terrorism, the United States is already conducting military operations in several other parts of the world. In the near future, these operations will probably include training, equipping, or providing

RANDMR1606-4.2

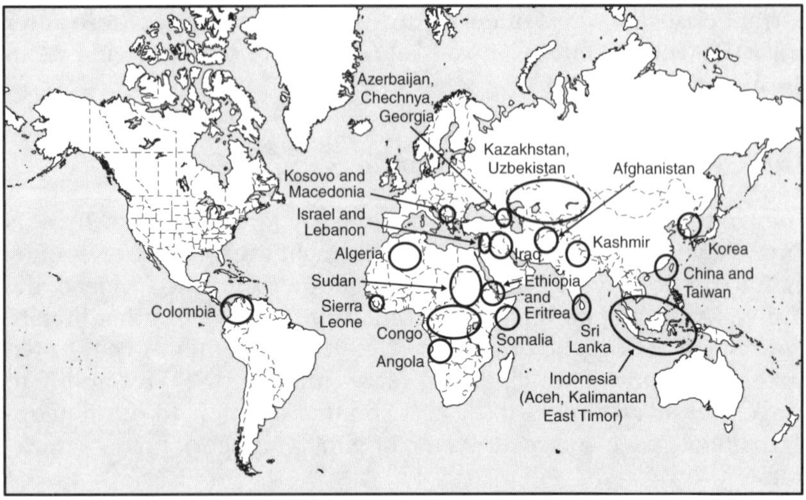

NOTE: For more information on such areas, see International Institute for Strategic Studies, *Strategic Survey 1999/2000*, London, 2000.

Figure 4.2—Areas of Current Crises and Conflicts

intelligence or logistics support to indigenous forces; special operations; and, perhaps, raids by company- or battalion-sized forces. They may also include brigade-sized operations, and possibly even full-scale invasion of states that sponsor or tolerate terrorist groups.

In consultation with the Attorney General and the Secretary of the Treasury and pursuant to the Immigration and Neutrality Act, the Secretary of State designates foreign terrorist organizations. At this writing, the Secretary of State has designated 28 such organizations, including al Qaeda, which financed and directed the attacks on September 11.[9] Al Qaeda began as an Arab resistance movement to the Soviet forces in Afghanistan. It may have several thousand members worldwide. It conducted terrorist attacks in Kenya, Tanzania, the Philippines, and Yemen prior to the September 11 at-

---

[9] U.S. Department of State, *2001 Report on Foreign Terrorist Organizations*, Washington, D.C.: Office of the Coordinator for Counterterrorism, October 5, 2001, available at http://www.state.gov/s/ct/ris/rpt/fto.

tacks in the United States. The goals of al Qaeda are expressed in "Jihad Against Jews and Crusaders" issued by Osama bin Laden and others on February 23, 1998.[10]

Following the Soviet withdrawal from Afghanistan, the country descended into civil war as resistance groups and tribal factions contended for power. Eventually, the Taliban, a movement that originated in mosque schools, took power in Kabul, but it was unable to conquer the northeastern part of the country, especially the Tajik-inhabited Panshir Valley. Al Qaeda enjoyed a sanctuary in Afghanistan under Taliban rule. After the Taliban leaders refused to crack down on al Qaeda following its attack on the United States, the United States conducted air strikes against Taliban government facilities, military barracks, air defenses, aircraft, and deployed forces. Air operations soon focused on support to the Northern Alliance, a loose coalition of anti-Taliban forces fighting since the Taliban took power. Close air support and other battlefield air operations directed by U.S. Special Operations Forces allowed the Northern Alliance to overrun Taliban positions. This success impelled several Pushtun-dominated groups in southern and central Afghanistan to oppose the Taliban, causing a rapid dissolution of its power. Meeting in Germany, Afghan delegates approved a provisional government, which assumed responsibility in Kabul, protected by an international peacekeeping force. The United States continues to hunt for al Qaeda members, especially the charismatic leader Osama bin Laden, who was presumed alive and possibly still in Afghanistan at the time this report was completed.

Deployment of operations against terrorist groups is a "wild card": Terrorists might hide almost anywhere. Thus far, U.S. interest has focused on those areas that promise sanctuary because the central governments exert little control and the populations, if not sympathetic, are at least Islamic. But in the future, al Qaeda might try to hide in very different parts of the world. They might, for example,

---

[10]"Jihad Against Jews and Crusaders" appeared in February 1998 in *Al-Quds al-Arabi*, a newspaper published in London. It purports to be a *fatwa* ("religious edict") issued by an organization calling itself the World Islamic Front. The signatories included Shaykh Usamah Bin-Muhammad Bin-Laden in a personal capacity and representatives of Islamic organizations in Bangladesh, Egypt, and Pakistan. The declaration has been widely reproduced and disseminated on the Internet. It is available at http://www.library.cornell.edu/colldev/mideast/wif.htm.

lead a clandestine existence in Western countries, including in the United States itself, where the perpetrators of the September 11 attack took flight training.

In addition to Afghanistan, the United States is currently conducting or preparing operations to hunt down al Qaeda members and to destroy groups associated with al Qaeda in Georgia, Somalia, the Philippines, and Yemen. Somalia and Yemen are predominantly Islamic countries. In Georgia and the Philippines, the areas of interest contain substantial Islamic populations. Thus, operations to date have focused on areas where al Qaeda might hope to find support from co-religionists. (See Figure 4.3.)

Georgia. Since the break-up of the Soviet Union, Georgia has suffered from internal lawlessness and a tense relationship with Russia. Russia has fought several times against separatists in Chechnya, a predominantly Islamic region that attempted to become independent from Russia.[11] In 2001, Russian President Vladimir V. Putin declared that the war was over in Chechnya, but Russian forces continue to conduct sweep operations. Some of the rebels and numerous refugees fled from Chechnya through the rugged Caucasus Mountains into northeastern Georgia, especially into the Pankisi Gorge, which Georgia's government does not control. Georgia remained neutral during the Chechen war and has refused to assist Russia against the rebels, but it ultimately needs to regain control of its territory. The government of the United States believes that the rebels and outlaws who frequent the Pankisi Gorge have connections to al Qaeda, and it plans to equip four Georgian battalions and to train them for at least six months. The Russian government, which still has thousands of troops stationed in Georgia, may regard this initiative with some skepticism; however, President Putin announced: "We support this fight no matter who takes part in it."[12]

---

[11]See Olga Oliker, *Russia's Chechen Wars 1994–2000: Lessons from Urban Combat,* Santa Monica, Calif.: RAND, MR-1289-A, 2001.

[12]Sharon LaFraniere, "U.S. to Train 1,200 Georgian Soldiers in Terror Battle," *The Washington Post,* March 2, 2002, pp. A14, A20.

RANDMR1606-4.3

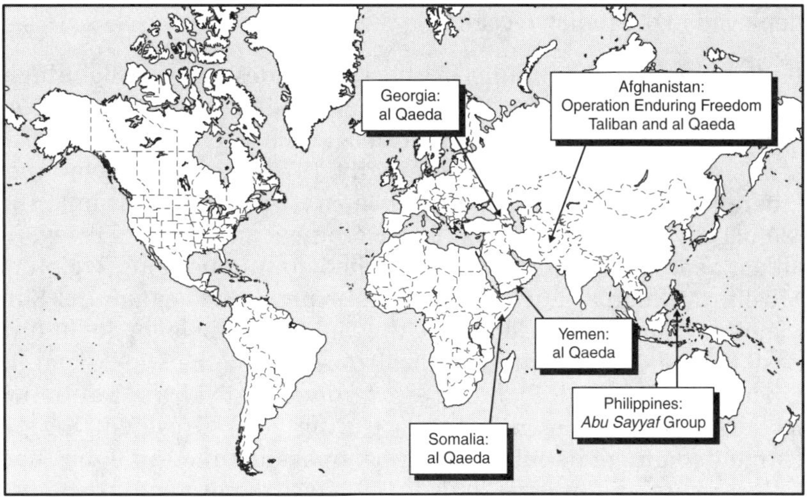

Figure 4.3—Efforts Against Terrorist Groups

Somalia. Osama bin Laden claimed to have assisted Somali clans against U.S. forces during the conflict that culminated in a protracted firefight in Mogadishu on October 3–4, 1993. The incident prompted the Clinton administration to withdraw U.S. forces from Somalia. Al Qaeda used bases in Kenya and Somalia to prepare their bombing attacks on U.S. Embassies in Dar es Salaam and Nairobi in 1998. The current interim government of Somalia does not control large areas of the country. Following the September 11 attack, Somalia's Prime Minister Hassan Abshir Farah, leading the Transitional National Government, said that his government supported the international effort against terrorism and would welcome deployment of U.S. forces to track down terrorists in Somalia. The United States, France, and the United Kingdom are currently flying reconnaissance over Somalia, including U.S. Navy P-3 aircraft flying out of Oman.[13] In the near future, the United States might conduct special operations in

[13]Thomas E. Ricks, "Allies Step Up Somalia Watch, U.S. Aims to Keep al Qaeda at Bay," *The Washington Post,* January 4, 2002a, p. A1.

Somalia or employ larger forces, such as a Marine Expeditionary Unit deployed in the Indian Ocean.

Philippines. The Republic of the Philippines is overwhelmingly Roman Catholic, but a Muslim minority lives on the island of Mindanao and on the Sulu Archipelago, which extends in an arc from Mindanao to Malaysia. In the early 1970s, the Moro National Liberation Front (MNLF) began a separatist rebellion against the Manila government. In 1996, the government and the MNLF negotiated a peace settlement that established an Autonomous Region of Muslim Mindanao encompassing four provinces: Lanao del Sur, Maguindanao, Tawi-tawi, and the Sulu Archipelago, including Basilan and Jolo. A more radical movement, the Moro Islamic Liberation Front (MILF), broke away from the MNLF and continues to confront government forces. During 2000, President Joseph Estrada conducted a military campaign against the MILF, but had little success. In January 2001, Estrada was deposed amid popular protest over corruption in his administration. Vice President Gloria Macapagal-Arroyo reversed Estrada's policy and concluded a cease-fire with the MILF. That cease-fire is still in effect, but there have been numerous violations.

In 1991, the *Abu Sayyaf* ("Bearer of the Sword") Group (ASG) began a violent campaign, kidnapping and assassinating Christian Filipinos. In 1995, ASG allegedly massacred some 50 villagers in the largely Christian village of Ipil. In April 2000, it kidnapped 21 people from a resort on the island of Sipidan in Malaysia. The government rejected the ASG's demand to exchange the Filipino hostages for Ramzi Yousef, convicted of plotting the 1993 bombing of the World Trade Center in New York. In May 2001, ASG kidnapped three Americans and 17 Filipinos from a resort off Palawan Island and held them for ransom. Of the Americans, one was murdered, one died during a rescue attempt, and one (Gracia Burnham) survived her ordeal. ASG probably can muster about 500 combatants equipped with small arms, machine guns, grenade launchers, mortars, and watercraft.[14]

---

[14]See BBC News, "Guide to Philippines Conflict," December 6, 2001, available at http://new.bbc.co.uk/hi/english/world/asia-pacific/newsid_169500/1695576.stm; Naval Postgraduate School, "Terrorist Group Profiles," Monterey, Calif.: Dudley Knox Library, available at http://web.nps.navy.mil/~library/tgp/asc.htm; Paul Wiseman, "Big Target Pinned on Small Rebel Group," *USA Today*, February 21, 2002, p. 4, avail-

In November 2001, to commemorate the 50th anniversary of the U.S.–Philippine Mutual Defense Treaty, Presidents George W. Bush and Macapagal-Arroyo met in Washington and issued a joint statement emphasizing their robust defense partnership in view of the September 11 attacks on the United States and the terrorist activities of the ASG. The United States has since increased military aid to the Philippines and has deployed forces to assist Philippine forces in fighting the ASG. U.S. Special Forces accompany Philippine forces, such as the Army Scout Rangers, patrolling on Basilan Island, but are not allowed to enter combat except in self-defense. U.S. forces also helped establish an elite Light Reaction Company specializing in hostage rescue, which is currently operating out of Isabela on Basilan Island.[15] The terrain on Basilan includes thick jungle, which restricts visibility to a few yards and presents an opaque canopy from above. U.S. forces are currently constructing all-weather roads linking Isabela City with other towns on the island, repairing the Isabela airstrip, and constructing fresh-water wells.[16]

Yemen. Osama bin Laden's family emigrated from Yemen to Saudi Arabia, where his father made an immense fortune as a contractor for large building projects. Members of al Qaeda are currently lurking in Yemen, especially in the north-central highlands, which are largely controlled by tribal groups. The Yemeni government called for cooperation in hunting these terrorists and warned against hiding them. During December 2001, the government attacked a tribe suspected of hiding al Qaeda members in the Mar'ib area east of the capital city of Sanaa. At the time of this writing, the United States and Yemen are discussing military assistance to Yemeni forces, including advisers.[17]

---

able at http://ebird.dtic.mil/Feb2002/e20020221big.htm; and Raymond Bonner, "U.S. and Philippine Governments Revive Old Relationship," *The New York Times*, March 4, 2002a, available at http://ebird.dtic.mil/Mar2002/e/e00e0304usand.htm.

[15]Jane Perlez, "Philippine Troops Eagerly Await U.S. Help and Arms," *The New York Times*, February 12, 2002, p. 1.

[16]Linda D. Kozaryn, "Attacks Continue in Afghanistan; Philippines Training Under Way," *Armed Forces Information Service*, April 24, 2002; Faber Conception, "Akbar Gets U.S. Assurance on Projects," *Manila Times*, April 30, 2002.

[17]Thomas E. Ricks, "U.S. Eyes Military Assistance for Yemen," *The Washington Post*, February 28, 2002b, p. 1; Vernon Loeb and Peter Slevin, "U.S. Special Forces to Advise Yemeni Military," *The Washington Post*, March 2, 2002, p. A20.

Terrorism: Looking to the Future

As has already been mentioned, terrorism is a "wild card": It might prompt the United States to use military force at short notice in unexpected places. Al Qaeda, for example, might operate clandestinely almost anywhere in the world, including the United States. It might also find refuge in existing insurgent movements, terrorist groups, and criminal conspiracies, including some that have no Islamic sympathies. But other groups may hesitate to welcome al Qaeda if they believe that the association would compel the United States to support their enemies.

REGIONS OF INTEREST

Past U.S. military operations, current areas of crisis and conflict, the locales of UN peace operations, and the pattern of state-sponsored terrorism point generally toward five regions of the world where U.S. forces are likely to operate (see Figure 4.4):

- Central America and the Caribbean
- Europe
- Northern and sub-Saharan Africa
- Middle East and Southwest Asia
- East Asia and the Pacific Rim.

Each of these regions has a distinctly different character. Central America and the Caribbean comprise a region of traditional interest in which the United States has repeatedly intervened, usually in an attempt to improve governance. Europe, also a region of traditional interest, is where the United States fought two major wars and, to protect it, faced off against the Soviet Union for 50 years. Northern and sub-Saharan Africa could see future U.S. operations, as in Somalia and Rwanda, although recent administrations have shown great reluctance to employ force here. In the Middle East and Southwest Asia, two very strong U.S. interests are at stake: survival of Israel and access to the world's largest proven oil reserves. The United States has fought its most protracted and sanguinary wars

RANDMR1606-4.4

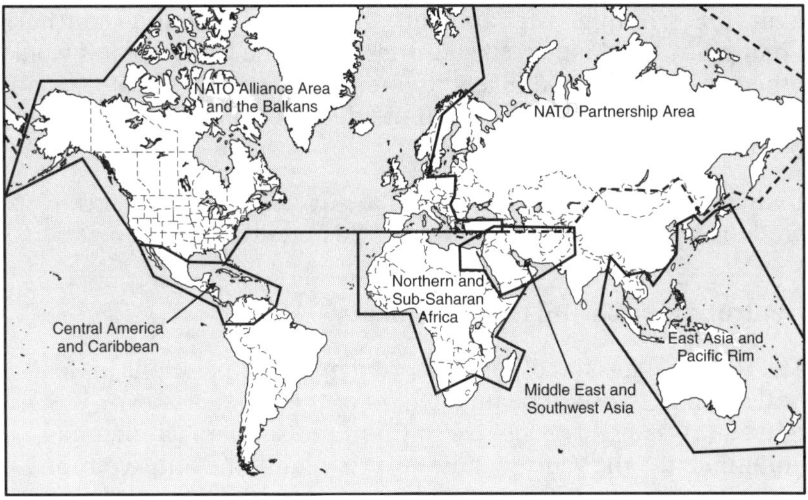

Figure 4.4—Regions of U.S. Operations

since 1945 in East Asia and the Pacific Rim. Moreover, the Taiwan issue poses the single greatest risk of conflict with another great power.

Although highly important, interests are not the sole determinants of U.S. policy abroad. The United States is founded on principles expressed in documents, such as the Declaration of Independence, the Gettysburg Address, and the Atlantic Charter, principles that include the equality of all people, their fundamental rights, and their freedom to govern themselves and that are universal, and hold as true in Kabul as in New York. Americans believe that their country has the historical destiny to realize these principles at home and to advance them abroad. But when the United States acts in the world, it consults its interests and recognizes the limits of its power, acting in these respects like any other country. As a result, the United States is in constant tension between its boundless principles and its narrower interests.

The terrorism wild card could draw the United States into regions where it otherwise would have little interest. For example, al Qaeda has thus far established bases where the majority of the population is

Muslim and the government is either weak or sympathetic to extremism—for example, Afghanistan, Indonesia, Somalia, southern Philippines, and Yemen, all countries where the United States would otherwise have been unlikely to operate. Moreover, as al Qaeda is driven from its accustomed haunts, it might appear in surprising places.

In this section, we examine each of the five regions for past U.S. operations, accessibility to operations, and potential future operations.

Central America and the Caribbean

The United States has often conducted military operations in Central America and the Caribbean, a region of traditionally strong U.S. interest. It was highly sensitive to Communist encroachment, as exemplified by the Cuban missile crisis and the intervention in Grenada. Events in the Caribbean are of particular interest to the United States when they generate flows of refugees, as has occurred from Cuba and Haiti.

Past Operations. During the Spanish-American War of 1898, the United States occupied Cuba and Puerto Rico. In 1904, it acquired rights in Panama to construct the Panama Canal. During the interwar years, the United States intervened forcefully in Cuba and Nicaragua. Following World War II, it sponsored an émigré invasion of Cuba (1961), quarantined Cuba to obtain removal of Soviet nuclear weapons (1962), assured an orderly transfer of power in the Dominican Republic (1965), overthrew a murderous Communist regime in Grenada (1983), forced a return to democratic government in Panama (1989), and ensured return of the legally elected government of Haiti (1994). It currently conducts counternarcotic operations in this region and supports efforts of the Colombian government to solve the related problems of narcotics smuggling and civil conflict.

In 1979, the Sandinistas, named after Nicaraguan revolutionary César Augusto Sandino, seized power in Nicaragua. They quickly expropriated large private holdings and established ties with Cuba and the Soviet Union. Starting in 1982, the Reagan administration funded the Nicaraguan Democratic Resistance ("contras") opposed to the Sandinistas. Congressional Democrats viewed the "contras" with great skepticism and passed legislation to prohibit or limit funding. Clandestine efforts to circumvent congressional limits caused the Iran-Contra scandal, which became public knowledge in November 1986. In 1990, the Sandinistas left office after losing a general election.

Accessibility. This region lies near the continental United States and is relatively shallow (see Figure 4.5). U.S. transport aircraft can reach every point in the region from CONUS without intermediate basing or aerial refueling. For example, Bogota, Colombia, is approximately 1,890 nmi from Pope AFB, North Carolina, which is collocated with the XVIII Airborne Corps and the 82nd Airborne Division. Moreover, except for the interiors of Mexico, Colombia, and Venezuela, areas of interest lie close to littorals. As a result, the United States has experienced little difficulty mounting rapid, decisive operations throughout the region. In countries such as Haiti, Grenada, and Panama, the United States achieved decisive results in a few days.

RANDMR1606-4.5

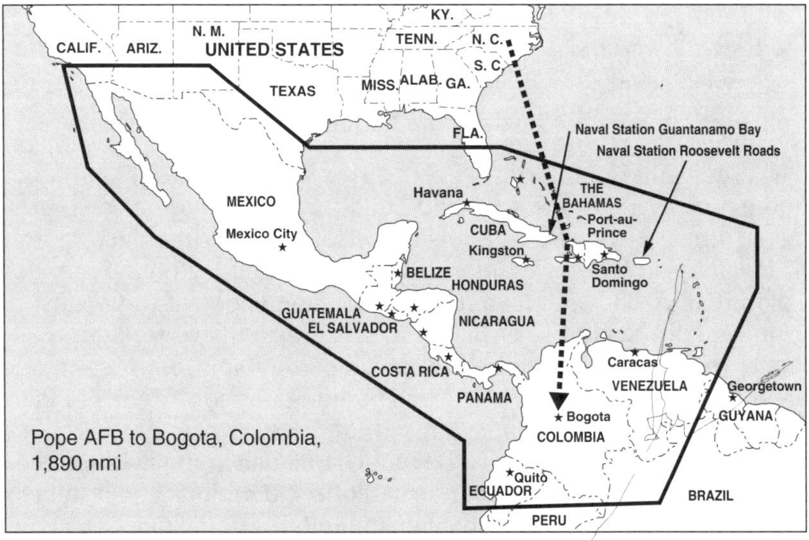

Figure 4.5—Central America and the Caribbean

Future Operations. The United States will probably conduct military operations in Central America and the Caribbean again, possibly intervening (in the pattern of past operations) to promote democratic governance and to avert the consequences of despotic governance, including flows of refugees such as the Haitian "boat people." For example, it might have to intervene in Cuba if the country were to implode after the demise of its Communist government.

At this writing, Colombia poses perhaps the most difficult problem for U.S. security policy in the Western hemisphere. Political violence and rebellion have plagued this country almost continuously for the past four decades. The administration of Andres Pastrana came into office pledging to seek a solution through negotiation. In November 1998, the Pastrana government withdrew from some 42,000 square kilometers of territory in south-central Colombia, thus ceding control to the *Fuerzas Armadas Revolucionarias de Colombia* ("Revolutionary Armed Forces of Colombia," FARC), which had made government withdrawal a precondition for negotiations.

Well financed through drug trafficking, extortion, and kidnapping, the FARC is a Marxist revolutionary movement that currently controls 15,000–20,000 combatants organized in small units of light infantry force, with uniforms and formal rank structure. The situation is complicated by the growing strength of the *Ejercito de Liberación* ("National Liberation Army," ELN), a rival insurgent group operating in the northern part of the country and that may obtain its own protected area. Illegal self-defense groups, loosely organized as the *Autodefensas Unidas de Colombia* ("United Self-Defense Forces of Colombia," AUC), operate outside government control and are using brutal tactics against the insurgents. With extensive foreign assistance, the Pastrana government developed a Plan Colombia to address all aspects of the crisis. The United States contributes to Plan Colombia, especially through support of three new counternarcotics battalions equipped with UH-1H and UH-60 helicopters. U.S. forces train Colombian forces in counternarcotics operations and also help conduct surveillance directed against narcotics trafficking. The government is committed to negotiating with the rebels, but they have shown no willingness to disarm so long as they appear to be gaining ground. As a result, the Colombian crisis is likely to endure and provoke a stronger U.S. response.[18]

The United States is unlikely to mount large joint operations in countries below the Caribbean littoral. Maintaining close military ties to countries in South America and supporting regional peacekeeping, the United States has not conducted military operations in this part of the continent. Lack of direct interests, aversion of governments to U.S. intervention, and distance all make future operations here less probable. That said, if global terrorist cells are discovered in South America or governments begin to develop WMD, the prospect for major military operations would rise significantly.

Europe

The United States has strong historical interests in Europe, fighting in both World Wars I and II to protect those interests and basing large forces in Europe during the almost 50 years of the Cold War. It was willing to risk nuclear attack on the U.S. homeland to defend Europe, which today appears largely secure from conventional aggression. However, the United States could be drawn into operations on the periphery as it was in Bosnia and Kosovo.

Past Operations. During the Cold War, the United States and its NATO allies confronted the Soviet Union and Warsaw Pact countries in Western Europe. Perhaps the most important military operation conducted during this confrontation was Operation Vittles, the Berlin Airlift, from June 1948 to September 1949. The United States initially tried to remain aloof from the Balkan wars that followed Yugoslavia's dissolution, eventually concluding that its leadership was indispensable. In a typical pattern, NATO conducted air efforts (Deny Flight, Deliberate Force, Allied Force) before undertaking enforcement operations (Joint Endeavor, Joint Guardian).

During the Bosnian war, the United States enforced a no-fly zone (Deny Flight), provided close air support to the United Nations Protection Force (UNPROFOR), and contributed to NATO operations

---

[18]For a recent analysis of the Colombian crisis, see Angel Rabasa and Peter Chalk, *Colombian Labyrinth: The Synergy of Drugs and Insurgency and Its Implications for Regional Stability*, Santa Monica, Calif.: RAND, MR-1339-AF, 2001a.

in Bosnia (Joint Endeavor) and Kosovo (Joint Guardian). It initially decided not to contribute troops to a NATO effort in Macedonia, but may eventually be drawn into a new operation there as well. The operations in Bosnia and Kosovo are open-ended. After six years of effort in Bosnia, the ostensible goal of harmonious multi-ethnic government is still far from attainment. NATO thus appears destined to continue modest-sized peace operations in the Balkans indefinitely.

Accessibility. Transport aircraft flying from the United States require intermediate stops or aerial refueling to reach the territory of the NATO Alliance members in Western Europe. For example, Berlin, Germany, is approximately 3,860 nmi from Pope AFB (see Figure 4.6). The United States maintains air bases in England, Germany, Italy, and Turkey, and it enjoys access to bases in other member countries. However, in all contingencies outside Article 5,[19] members grant or withhold basing rights on a case-by-case basis. For example, members other than Great Britain declined to allow use of their airspace during El Dorado Canyon, a punitive air operation against Libya during the Reagan administration.

The Seventh Army is stationed in Germany, the 1st Armored Division is headquartered in Bad Kreuznach, and the 1st Infantry Division is headquartered in Würzburg. Each of these divisions has one brigade stationed in the United States. Army War Reserve (AWR)-2 includes brigade sets stored in Belgium, Luxembourg, and the Netherlands. MPS Squadron One is usually located in the Mediterranean Sea. The Marine Corps maintains an equipment set in Norway for the Norway

---

[19]The North Atlantic Treaty, Washington, D.C., April 4, 1949: "Article 5: The Parties agree that an armed attack against one or more of them in Europe or North America shall be considered an attack against them all and consequently they agree that, if such an attack occurs, each of them . . . will assist the Part or Parties so attacked by taking forthwith, individually and in concert with the other Parties, such action as it deems necessary, including the use of armed force. . . ." Article 5 directly obligates each member only to "such action as it deems necessary" but it is taken to imply a commitment to collective defense consistent with the constitutional provisions of each member country. On September 12, 2001, the North Atlantic Council voted unanimously to invoke Article 5 if it was determined that the previous day's attacks on the United States were directed from abroad. Upon determination that the attacks were directed by al Qaeda under leadership of Osama bin Laden, Article 5 was invoked for the first time in NATO history.

RANDMR1606-4.6

Figure 4.6—North Atlantic Treaty Organization

Air-landed Marine Expeditionary Brigade (NALMEB) under terms of a 1974 agreement. In addition, the Army has equipment for one artillery battalion stored in Norway.

Future Operations. In the near term, it is hard to imagine any U.S. military operation in Europe that would not be either under the auspices of NATO or done with the blessing of key member states. For this reason, the following discussion focuses on the changing nature of the NATO Alliance.

Originally founded as an alliance for collective defense, embodied in Article 5 of the Washington Treaty, NATO after the Cold War directed its attention increasingly toward non–Article 5 ("out of area") conflicts, especially the Balkan wars. At the same time, it began an enlargement program.[20] Thus far, NATO has added just three new members: Poland, Hungary, and the Czech Republic. In 1999, NATO launched its Membership Action Plan (MAP) to prepare certain countries for membership. More broadly, the alliance invited nonmembers to join the Euro-Atlantic Partnership Council (EAPC) and to participate in Partnership for Peace. Twenty-seven countries are now members of the EAPC.[21]

One effect of NATO enlargement is to blur the line between those countries within and outside the alliance and those countries outside it. Previously, there was a fairly distinct line between alliance members, who collectively guaranteed each other's security, and nonmembers, who did not enjoy such a guarantee. Today, the guarantee would almost certainly extend in practice to countries that are outside the alliance but are members of the European Community and the Partnership for Peace.

On 10–11 January 1994, NATO's North Atlantic Council (NAC) meeting in Brussels at the level of heads of state announced the establishment of Partnership for Peace (PfP). The NAC linked PfP to the evolutionary expansion of the alliance, in effect making PfP an initial step toward membership in NATO. The NAC committed itself to consult with any participant in PfP that perceived a direct threat to its territorial integrity, independence, or security. In an annex to the communiqué (M-1[94]2), 10–11 January 1994), the NAC required that all PfP members reaffirm their commitment to the Charter of the United Nations, the Universal Declaration on Human Rights, and the Helsinki Final Act. PfP members must also agree to cooperate with NATO in pursuing transparency in defense spending, democratic control of forces, capability to contribute to operations under authority of the UN or the Organization for Security and Cooperation in

---

[20] For an analysis of NATO enlargement culminating in the Madrid Declaration on Euro-Atlantic Security and Cooperation issued by Heads of State and Government meeting in the North Atlantic Council on July 8, 1997, see Gerald B. Solomon, *The NATO Enlargement Debate 1990–1997: The Blessings of Liberty*, Westport, Conn.: Praeger, 1998.

[21] See Thomas S. Szayna, *NATO Enlargement 2000–2015: Determinants and Implications for Defense Planning and Shaping*, Santa Monica, Calif.: RAND, MR-1243-AF, 2001, pp. 49–72.

Europe (OSCE), cooperative militant relations with NATO, and development of forces better able to operate with NATO forces.

European security arrangements might be visualized as concentric circles, radiating from the members of the NAC, to members of the European Union (who are all members of either the NAC or the EAPC), to the MAP countries, to other countries in the EAPC, and finally to the countries of the OSCE. Each larger circle would enclose an area of decreasing security commitment.[22]

The Organization for Security and Cooperation in Europe is a regional security organization active in early warning, conflict prevention, crisis management, and post-conflict rehabilitation. It also has a broad charter to promote human rights. It traces its origin to the Final Act of the Conference on Security and Cooperation in Europe, signed on August 1, 1975, in Helsinki, Finland, and is commonly known as the Helsinki Final Act.

NATO faces very little risk to its members from outside the alliance. Indeed, it faces a greater risk within the alliance if Greece and Turkey fight over Cyprus. NATO might undertake additional non–Article 5 operations. But for the foreseeable future, it probably will be preoccupied with apparently interminable operations in Bosnia, Kosovo, and perhaps Macedonia as well. If NATO were to undertake additional non–Article 5 operations, they might occur in the area from Turkey to the southern perimeter of the former Soviet Union, including the countries of the Caucasus (Armenia, Azerbaijan, Georgia) and Central Asia (Kazakhstan, Kirgiz Republic, Tajikistan, Turkmenistan, Uzbekistan).

Russia regards the Caucasus and Central Asia as a traditional sphere of influence and has tried to promote a Commonwealth of Independent States (CIS) under its leadership. Fears of Russian domination have caused the CIS to cease to be an effective vehicle for economic or military cooperation. If NATO were to undertake independent operations in this area, it would have to expect a very negative reaction from Russia. Apart from Russia itself, the countries in this area are threatened more by internal instability than by foreign invasion—and that threat appears remote. The oil-producing

[22]Szayna, 2001, pp. 30–36.

countries are currently trying to encourage foreign investment to exploit their reserves more effectively. These reserves are very large. Kazakhstan, for example, could become one of the world's largest oil producers and exporters in the next decade.[23] In early 2001, the Caspian Pipeline Consortium began filling a pipeline connecting Kazakhstan's huge Tengiz oil field with a marine terminal on the Black Sea. NATO might conduct humanitarian operations, enforce sanctions, and perhaps even contribute troops to peace operations under other auspices, but appears unlikely to intervene forcibly in this region:

> Inasmuch as the Alliance lacks the collective interest, will, capabilities, and resources to assume responsibility for Caspian security, a U.S.-led coalition of willing countries rather than NATO should assume primary responsibility for securing Western objectives in the Caspian basin.[24]

Northern and Sub-Saharan Africa

Historically, the United States has not viewed Africa as a security priority. Humanitarian and human rights concerns have led to some involvement (e.g., support for peace operations, intervention in Somalia, and past sanctions against South Africa), and instability has forced the United States to conduct noncombatant evacuation operations in several countries. Since World War II, military planners have not seriously contemplated conducting large-scale operations on the continent. As the United States recognizes the African role in global terrorism, this may change. Somalia, Sudan, and Libya have all provided sanctuary for terrorists in the past, and smuggling of diamonds out of the Congo is increasingly recognized as a source of funding for terrorist organizations.[25] Somalia, the Congo, and Sierra Leone are failed states that are likely to continue to provide havens for international criminal and terrorist groups. Isolated raids against

---

[23]See analysis by the U.S. Department of Energy, available online at http://www.eia.doe.gov/emeu/cabs/kazak.html.

[24]Richard Sokolsky and Tanya Charlick-Paley, *NATO and Caspian Security: A Mission Too Far?* Santa Monica, Calif.: RAND, MR-1074-AF, 1999, p. 98.

[25]Douglas Farah, "Digging Up Congo's Dirty Gems," *The Washington Post,* December 30, 2001, pp. A1, A16.

terrorist groups in these countries may occur as the United States seeks to disrupt their activities. The United States might also become involved in more-ambitious and more-aggressive initiatives to restore stability to these nations.[26]

Past Operations. Since World War II, the United States has generally avoided operating in Africa, with the notable exception of Somalia.[27] The humanitarian intervention (Restore Hope) was successful, but the subsequent attempt to implement the Addis Ababa Accords (Continue Hope) was badly bungled and ended disastrously. Finally, the United States had to conduct a military operation (United Shield) to extract the hapless UN force from Somalia. The failure in Somalia seems to have been a major cause of U.S. reluctance to operate in Rwanda, even as evidence of genocide mounted. In contrast to its Balkan policy, the United States has not contributed military forces to peace operations in other African countries such as Angola, the Democratic Republic of the Congo, Liberia, Mozambique, and Sierra Leone. Instead, it supports peace operations conducted by other countries. For example, Focus Relief will train and equip up to seven battalions from West African countries for peace operations.

Accessibility. Africa lies far from CONUS and has enormous extent (see Figure 4.7). For example, the distance from Pope AFB to Kigali, Rwanda, is approximately 6,400 nmi, and the distance from Ramstein Air Base in Germany to Kigali is 3,500 nmi. Therefore, intermediate staging or in-flight refueling would be required in order to deliver substantial payloads. These distances, coupled with low capacities at receiving airports, make airlift challenging.

Future Operations. Interstate wars, civil wars, and ethnic or clan-based conflict will cause numerous countries in northern and sub-Saharan Africa to remain unstable. Ethiopia and Eritrea appear to have concluded a protracted border war. Angola, the Democratic

---

[26]See Steven Metz, *Refining American Security in Africa*, Carlisle, Penn.: U.S. Army War College, Strategic Studies Institute, 2000, p. 3.

[27]After securing delivery of humanitarian aid, the United States attempted to capture the clan leader Mohhamed Farah Aideed, leading to the firefight depicted by Mark Bowden in *Black Hawk Down*, New York: Atlantic Monthly Press, 1999. Thereafter, the United States assumed a defensive posture and withdrew its forces from Somalia.

RANDMR1606-4.7

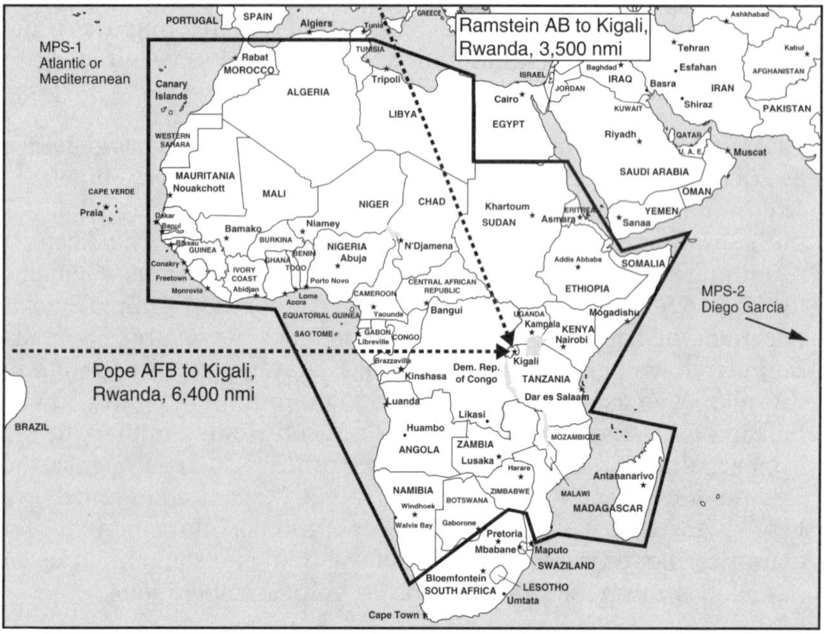

Figure 4.7—Northern and Sub-Saharan Africa

Republic of the Congo, and Sudan are suffering from protracted civil wars. The involvement of neighboring countries has exacerbated the problem in the Congo. Algeria is suffering from a fundamentalist Muslim rebellion that resorts to terrorism. Somalia, especially the Mogadishu area, remains in a state of constant friction among clan-based groups. Ethnic violence between Hutus and Tutsis could become acute in Burundi or Rwanda with little warning. These conflicts and others are certain to cause humanitarian emergencies and to occasion UN operations such as those currently under way in the Congo, Eritrea-Ethiopia, Sierra Leone, and the western Sahara. To the extent that the prevalence of conflict may inhibit rather than provoke a U.S. response makes the region's problems appear intractable. The United States will help provide humanitarian relief, as it did in Mozambique (Operation Atlas Response, 2000); it will evacuate its nationals when they are endangered, as it did recently in Liberia (Operation Assured Response, 1996); and it may well conduct operations against terrorist groups, but it seems unlikely that it will conduct large military operations.

Middle East and Southwest Asia

The United States is closely engaged in this region to ensure the survival of Israel and maintain access to oil reserves. These two interests coincided during the 1990–1991 Persian Gulf War when Saddam Hussein seized Kuwait for the avowed (if little-credited) purpose of overwhelming Israel. To make the connection more plausible, he attacked Israel with ballistic missiles during the war.

Past Operations. The United States landed troops in Lebanon in 1958 (Operation Blue Bat) and 1982–1983 (Multinational Force) in attempts to stabilize that country. In October 1983, a terrorist truck bomb destroyed the Marine barracks near Beirut, killing 241 Marines and causing the United States to withdraw its forces. During the Yom Kippur War, the United States airlifted military assistance to Israel (Operation Nickel Grass), and it still contributes troops to a monitoring mission in the Sinai established after that war. Following Iraq's seizure of Kuwait, the United States enforced sanctions (Maritime Intercept Operations), defended Saudi Arabia (Operation Desert Shield), and eventually liberated Kuwait (Operation Desert Storm) during 1990–1991. During the Persian Gulf War, it deployed Patriot batteries to help defend Israel against Iraqi ballistic missiles, and later in 1991 with its coalition partners, the United States provided humanitarian assistance to the Kurdish population of northern Iraq (Operation Provide Comfort I). In response to threatening Iraqi deployments during October 1994, the United States deployed additional forces to Kuwait (Operation Vigilant Warrior). In 1996, it conducted cruise-missile strikes (Operation Desert Strike) in response to an Iraqi attack on Kurds, and in 1998, conducted air attacks (Operation Desert Fox) to compel Iraqi compliance with the United Nations Special Commission (UNSCOM) on Iraq, which was trying to investigate Iraq's programs to develop weapons of mass destruction. Currently, the United States is enforcing no-fly zones encompassing much of Iraqi airspace (Operations Northern and Southern

RANDMR1606-4.8

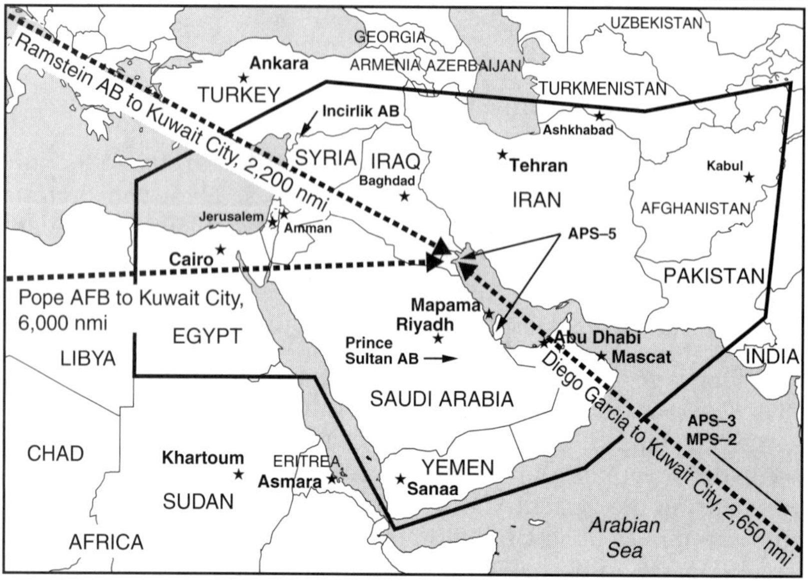

Figure 4.8—Middle East and Southwest Asia

Watch).[28] On October 7, 2001, President George W. Bush announced the start of strikes against al Qaeda training camps and military installations of the Taliban regime in Afghanistan (Operation Enduring Freedom).

Accessibility. This region is at the farthest remove from CONUS and difficult to reach even from U.S. bases in Western Europe (see Figure 4.8). Transport aircraft flying from the United States need at least two intermediate stops or else aerial refueling to reach this region. Although Kuwait City is 2,200 nmi from Ramstein AB, a straight-line flight would cross Syrian and Iraqi airspace. Avoiding this airspace demands intermediate stops or refueling for C-17 and C-5A/B aircraft flying with useful payloads. To mitigate these great distances, the United States maintains brigade equipment sets in Kuwait and Qatar, designated AWR-5. In an emergency, the United States might

[28]Operation Northern Watch was originally named Operation Provide Comfort II.

also deploy a brigade set afloat AWR-3, as it did in response to Iraqi deployments in 1994 (Operation Vigilant Warrior). MPS Squadron Two is usually located at Diego Garcia.

Future Operations. Most countries of the Middle East and Southwest Asia have authoritarian governments. The exceptions are Israel, which is democratic, and Iran, which holds free elections but is still heavily influenced by its Islamic revolution. Although the undemocratic regimes have been remarkably stable, they may ultimately prove unable to cope with rapid population growth and stagnant economies characteristic of most countries in the region, including some major oil exporters. At the same time, Islamic fundamentalism continues to rise, exacerbating Arab-Israeli tensions. Israel is embroiled in an unending quarrel with the Palestinian Arabs, who are deeply hostile to a Jewish state and may not be willing to coexist under any conditions that Israel could accept. On the positive side, Israel has not fought any of its neighboring states since 1982; Iraq is effectively contained; and Iran is starting to reduce the tensions with Gulf States caused by its strident fundamentalist principles. Even so, the entire region remains unstable and could erupt in conflict that would prompt the United States to conduct large military operations at short notice.

The United States has been preoccupied with containing Iraqi aggression since the Persian Gulf War. In the aftermath of that war, Saddam Hussein was fairly compliant: He removed forces from Kurdish areas and accepted UN inspectors charged with dismantling his programs to develop weapons of mass destruction. Subsequently, he has become more recalcitrant, particularly concerning weapons of mass destruction. By 1998, despite U.S. air strikes, Saddam Hussein had freed Iraq of intrusive inspections, allowing his regime to continue developing prohibited weapons. However, "without UNSCOM inspections, sanctions, and other measures Iraq would probably have a nuclear weapon and a range of biological weapons."[29] If the current Iraqi regime did acquire nuclear weapons, there would be grave consequences for Gulf security.

---

[29]Daniel L. Byman and Matthew C. Waxman, *Confronting Iraq: U.S. Policy and the Use of Force Since the Gulf War*, Santa Monica, Calif.: RAND, MR-1146-OSD, 2000, p. 74.

Since its Islamic revolution in 1979, Iran has been a bitter opponent of the United States, especially on a rhetorical level. Iran's cleric rulers are discredited, especially among younger Iranians, by the international isolation, domestic repression, and stagnant economy that their policies have caused. For years, Iranian politics have been almost paralyzed by infighting between radicals, who are motivated by the vision of a militant Islam, and moderates, who want to strengthen the economy through foreign investment, which entails a less militant foreign policy. Iran has assassinated opponents of the cleric regime abroad and has directly supported terrorism. It is developing medium-range ballistic missiles and may be attempting to develop nuclear warheads.

Iran may eventually become more accommodating to the West, but any Iranian government will probably see itself as a natural regional leader and, therefore, oppose Western influence to some degree. However, there is a fundamental cultural antagonism between Sunni Arabs and Shi'a Persians that will make Iran an implausible leader.[30] Even now, "Iran's ideology is often a mask for *Realpolitik*"[31] and its foreign policy is increasingly dictated by national interests.

East Asia and the Pacific Rim

The United States has vital interests in the Pacific Rim region and has been actively engaged there since the late nineteenth century. This engagement has been particularly beneficial in the past generation:

> Over the past 20 years, Asia has undergone a remarkable transformation. Under an umbrella provided by U.S. security guarantees and American military presence, the region has witnessed tremen-

---

[30]See Graham E. Fuller, *The "Center of the Universe": The Geopolitics of Iran*, Boulder, Colo.: Westview Press, 1991, pp. 8–31, 58–68; also see Graham E. Fuller and Bruce R. Pirnie, *Iran: Destabilizing Potential in the Persian Gulf*, Santa Monica, Calif.: RAND, MR-793-OSD, 1996, pp. 3–16.

[31]Daniel Byman, Shahram Chubin, Anoushiravan Ehteshami, and Jerrold D. Green, *Iran's Security Policy in the Post-Revolutionary Era*, Santa Monica, Calif.: RAND, MR-1320-OSD, 2001, p. 2.

dous economic growth, an expansion of its democratic institutions, and relative peace.[32]

Past Operations. By far the largest and costliest U.S. military operations since World War II were conducted in Korea and Vietnam. The conclusion of the Korean War, formally only an armistice, left the United States with an open-ended commitment to South Korean security. However, U.S. forces remain in Korea as much to promote regional security as to defend the peninsula against an invasion by North Korea. The conclusion of the Vietnam War was a Communist victory that completely excluded U.S. influence until very recently. The Vietnamese communists are also strongly nationalistic and deeply suspicious of Chinese power.

Accessibility. As Figure 4.9 illustrates, the Pacific Rim is almost as far from CONUS as is the Persian Gulf. However, the United States can use Hawaii to stage its forces and also has bases on Okinawa and Guam.[33] In contrast to the Japanese island of Okinawa, Guam and the nearby Northern Mariana Islands of Rota, Tinian, and Saipan are all U.S. territories, ensuring access in peacetime and crisis. Although the Northern Marianas do not currently host U.S. military facilities, they did during World War II and could again if Guam became overtaxed.

Most of the U.S. bases in East Asia and the Pacific Rim are positioned to secure South Korea. If U.S. forces had to protect Taiwan in some confrontation with China, basing could be a serious problem. Taiwan would almost certainly offer bases, but "it is difficult to imagine anything that would anger Beijing more than seeing U.S. forces arrive in Taiwan during a period of heightened

---

[32]Zalmay Khalilzad et al., *The United States and Asia: Toward a New U.S. Strategy and Force Posture*, Santa Monica, Calif.: RAND, MR-1315-AF, 2001, p. xi.

[33]The United States acquired Guam from Spain at the end of the Spanish-American War in 1898. Guam became an unincorporated territory in 1950, but currently seeks to improve its status under a proposed Guam Commonwealth Act. The United States acquired the Northern Mariana Islands in 1945 as a consequence of the Japanese surrender and initially administered the islands as a UN Trust Territory. In 1967, the U.S. Congress approved a negotiated covenant that established a Commonwealth of the Northern Mariana Islands in political union with the United States. Legally qualified residents of the Commonwealth are U.S. citizens, and a Resident Representative represents the Commonwealth before Congress and the federal government.

RANDMR1606-4.9

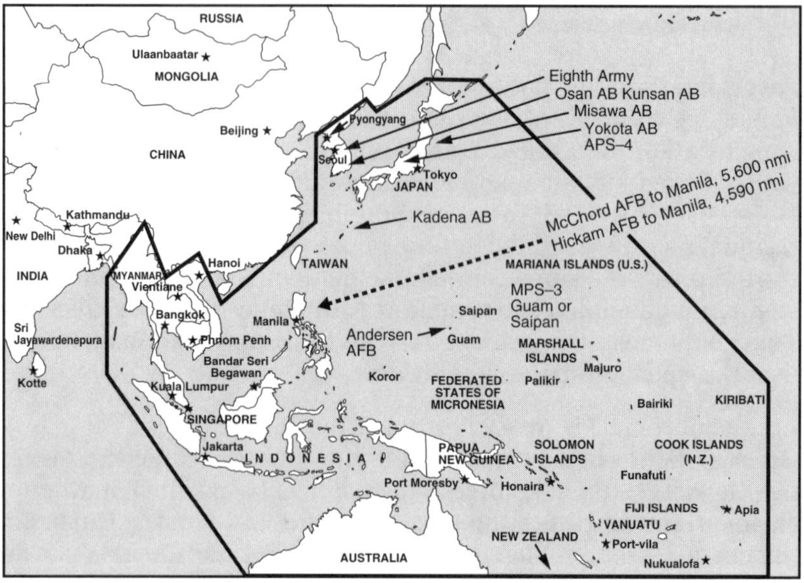

Figure 4.9—East Asia and the Pacific Rim

tensions. . . ."[34] But if basing in Taiwan were precluded for political reasons, the United States would have only one base within an unrefueled fighter radius (approximately 500 nmi) of Taiwan: Kadena Airbase in Okinawa, which currently supports only two fighter squadrons. Guam could serve as a staging base and logistics hub, but U.S. forces would require forward bases much closer to the area of operations. With Japan's permission, U.S. forces might base in the Southern Ryukyu Islands, which extend in a southwest arc from Okinawa toward the northern tip of Taiwan. The closest of these islands, Yonaguni, lies only 150 nmi east of Taipei. The

---

[34]Khalilzad et al., 2001, p. 69.

Philippines might also allow temporary basing rights during a crisis. Other countries in Southeast Asia would be likely to deny basing rights, largely because they would fear Chinese retaliation.[35]

Future Operations. From the perspective of U.S. strategy, East Asia and the Pacific Rim are predominantly an air and naval theater of operations. The United States is unlikely to undertake large-scale operations on the Asian mainland, apart from Korea; even then, the Republic of Korea would contribute most of the land forces. However, the United States might elect to employ smaller land forces in Asia with missions such as the following:

- Monitoring a zone of separation between Indian and Pakistani forces
- Demonstrating the U.S. commitment to the defense of Taiwan (unlikely)
- Helping halt a North Korean invasion
- Assisting the Philippine government in combating insurgency
- Leading efforts to allay ethnic conflict in Indonesia.

Each of these missions is discussed in turn.

India and Pakistan. When British India was partitioned in 1947, hundreds of thousands of people died in massive outbreaks of ethnic and political violence between Hindus and Muslims. After partition, India continued to govern the predominantly Islamic Jammu and Kashmir provinces. India and Pakistan have fought twice, and the two countries currently deploy forces along a line of confrontation through the disputed areas. Pakistan supports an independence movement within Kashmir; India conducts a counterinsurgency campaign. Moreover, ethnic tension between Hindus and Muslims in western India is increasing.

---

[35]Richard Sokolsky, Angel Rabasa, and C. R. Neu, *The Role of Southeast Asia in U.S. Strategy Toward China*, Santa Monica, Calif.: RAND, MR-1170-AF, 2000, pp. 64–65.

On May 11, 1998, India announced that it had tested three nuclear devices, including a thermonuclear device;[36] Pakistan quickly followed with its own tests. The two countries are probably building modest nuclear arsenals. India is much stronger conventionally and, therefore, may not feel compelled to keep its nuclear forces in high readiness. As the weaker power, Pakistan may think that it needs a more readily available deterrent. Both countries could deliver nuclear weapons by aircraft and are probably developing nuclear-armed ballistic missiles as well.[37]

If India and Pakistan went to war again, whether or not nuclear weapons were employed, the United States might participate in humanitarian operations or post-conflict peace operations—for example, to monitor a buffer zone.

China and Taiwan. China's future is perhaps the greatest uncertainty in East Asia. Since 1978, China has enjoyed unprecedented economic growth, but it remains to be seen whether economic growth will be accompanied by progress toward democracy. China is currently modernizing its antiquated military forces through foreign equipment purchases and domestic manufacturing. Within 10 to 15 years, China could contest control over contiguous seas and threaten U.S. regional bases with ballistic and cruise missiles. It could also defend its own airspace using mobile surface-to-air missiles and new fighter aircraft armed with sophisticated air-to-air missiles.[38]

The United States and China might clash if China were to blockade, attack, or invade Taiwan. For the Chinese, Taiwan is a highly emotional issue: "Growing up with pride in its civilization of a thousand years and sorrow about cruel humiliations, the Chinese consider

---

[36]See Ashley J. Tellis, *India's Emerging Nuclear Posture: Between Recessed Deterrent and Ready Arsenal*, Santa Monica, Calif.: RAND, MR-1127-AF, 2001, pp. 500–522.

[37]India has successfully developed two missiles suitable for delivering nuclear weapons: Prithvi-II and Agni-II. Agni-II is a two-stage solid-propellant ballistic missile tested to ranges over 2,000 km, and India is reportedly working on Agni-III, which could reach targets deep within China. Pakistan has Shaheen-II, claimed to have a 2,000-km range. In addition, Pakistan has announced work on a longer-range missile designated Ghaznavi.

[38]See Zalmay Khalilzad et al., *The United States and a Rising China: Strategic and Military Implications*, Santa Monica, Calif.: RAND, MR-1082-AF, 1999, pp. 77–83.

unification with Taiwan as the final milestone to ending its 'Century of Shame.'"[39] For the United States, defense of Taiwan is a moral and political imperative formalized in the Taiwan Relations Act of 1979, which states: "It is the policy of the United States . . . to consider any effort to determine the future of Taiwan by other than peaceful means including by boycotts or embargoes, a threat to the peace and security of the Western Pacific area and of grave concern to the United States. . . ."[40]

From a U.S. perspective, the defense of Taiwan is primarily an air and naval problem, not one likely to involve large land forces. However, the United States might consider deployment of land forces to Taiwan to stiffen defense and demonstrate commitment.

Korea. "How the division of the Korea Peninsula is resolved presents the greatest challenge not only to the United States, but to all of the countries in the region. There is no single security problem in Asia that could bring us into war with such certainty."[41] North Korea's industry is declining, and its agriculture is so weak that large numbers of people would starve without foreign assistance. Moreover, Pyongyang has lost support from Russia and China, which see greater advantage in good relations with Seoul. The Korean peninsula seems headed for a major crisis, but it is unclear whether such a crisis will result in the major theater war that has long been the focus of planning.[42] Indeed, the more populous and much further devel-

---

[39]Hui Wang, "U.S.–China Bonds and Tensions," in Shuxun Chen and Charles Wolf, Jr., eds., *China, the United States, and the Global Economy*, Santa Monica, Calif.: RAND, MR-1300-RC, 2001, Chapter Twelve, pp. 271–272. For various perspectives on the Strait issue, see Gerrit W. Gong, ed., *Taiwan Strait Dilemmas: China-Taiwan-U.S. Policies in the New Century*, Washington, D.C.: Center for Strategic and International Studies, 2000.

[40]Taiwan Relations Act of 1979, passed by Congress on March 29, 1979, and signed into law by President Jimmy Carter on April 10. The purpose of this Act was to establish a new relationship with Taiwan following U.S. recognition of the People's Republic of China as the sole legal government of China.

[41]Andrew Scobell and Larry M. Wortzel, *The Asia Pacific in the U.S. National Security Calculus for a New Millennium*, Carlisle, Penn.: U.S. Army War College, Strategic Studies Institute, 2000, p. 13.

[42]For an appraisal of possible scenarios, see Jonathan D. Pollack and Chung Min Lee, *Preparing for Korean Unification: Scenarios and Implications*, Santa Monica, Calif.: RAND, MR-1040-A, 1999.

oped Republic of Korea should be able to defend itself against a weakening threat from the north.

Philippines. The United States is currently assisting the Philippine government to combat the Islamic *Abu Sayyaf* Group, an extremely violent splinter group with connections to al Qaeda. The Philippine government is trying to negotiate a settlement with the much larger and well-established Moro Islamic Liberation Front. If negotiations broke down and the Philippine government confronted the MILF again, it would be likely to request U.S. assistance. However, opinion is divided on the issue of whether U.S. forces should be allowed to conduct combat operations.[43]

Indonesia. Indonesia is an archipelago comprising over 17,000 islands stretching some 5,000 km between the Indian and Pacific Oceans. Its population includes some 300 ethnic groups speaking 350 distinct languages. Indonesia faces several threats to its stability: a long-simmering rebellion in the oil-rich province of Aceh in northern Sumatra, strong separatist tendencies in mineral-rich West Papua (formerly known as Irian Jaya), large-scale ethnic and religious violence in the Moluccas, and horrific ethnic violence in Central Kalimantan on Borneo.[44] In addition, violent mobs have attacked the large Chinese population on several occasions. Some of the Islamic extremists in Indonesia apparently have connections to al Qaeda, but it is questionable whether the government would welcome overt U.S. assistance against them.

In January 1999, the Indonesian government unexpectedly announced that the East Timorese could decide for themselves whether they wished to be autonomous or independent. Autonomy would imply remaining within Indonesia, but with broad rights to local self-government. Independence would imply creation of a new sovereign state. When a popular consultation yielded an overwhelming vote against autonomy, East Timor erupted in violence between sepa-

---

[43]See Philip P. Pan, "Some Filipinos Say U.S. Presence May Fuel Rebel Support," *The Washington Post*, February 7, 2002, p. 13; Agence France-Presse, "Gun-Toting U.S. Soldiers Have Apologized, Says Macapagal," *Manila Times*, February 2002; Raymond Bonner, "'Yankee Don't Go' Is Message in Philippines Antiterror Rally," *The New York Times*, February 25, 2002b.

[44]Angel Rabasa and Peter Chalk, *Indonesia's Transformation and the Stability of Southeast Asia*, Santa Monica, Calif.: RAND, MR-1344-AF, 2001b, pp. 27–45.

ratist forces and pro-government militias, who conducted a reign of terror. Under heavy international pressure, Indonesia agreed to accept an international peace force, called the United Nations Transitional Authority in East Timor (UNTAET). Australia led this effort and contributed the largest contingent.[45]

Indonesia's huge population—fourth largest in the world—and its strategic position on key sea-lanes make Indonesia of critical importance to U.S. interests in Asia. Indonesia will certainly continue to suffer ethnic tensions and could even disintegrate violently. In some circumstances, the United States might lead a peacekeeping operation in Indonesia, requiring a substantial commitment of light and medium-weight land forces.

SUMMARY

This brief review indicates that the United States is far more likely to operate in some regions than in others. The United States is very likely to operate in Central America and the Caribbean, a volatile region too close to be ignored. It is very likely to operate in the NATO Alliance area and the Balkans, and does so at the time of this writing. It is less likely to operate in the NATO Partnership area, apart from exercises. It is less likely to operate in northern and sub-Saharan Africa, barring a major shift in policy. It is very likely to operate in the Middle East and Southwest Asia. The United States currently has a brigade-sized force in Afghanistan and may invade Iraq in the near future. It might conduct operations in Southeast Asia and the Pacific Rim; it would, of course, respond to an invasion of the Republic of Korea; and it might respond again to threats against Taiwan. The United States is unlikely to operate in South America, Asiatic Russia, India, or mainland China.

Of the regions where the United States is likely to operate, two are reasonably easy to reach. The Caribbean is within a few hours' flying time. Even Colombia, should the United States change policy and decide to operate there, can be reached without refueling. The entire NATO Alliance area has well-developed infrastructure and,

---

[45]See Peter Chalk, *Australian Foreign and Defense Policy in the Wake of the 1999/2000 East Timor Intervention,* Santa Monica, Calif.: RAND, MR-1409-SRF, 2001.

moreover, the United States maintains a small corps in Germany. Two regions remain in which the United States is likely to operate but which are difficult to reach: the Middle East and Southwest Asia, and East Asia and the Pacific Rim. It follows that the United States can expect the greatest gains in strategic responsiveness by focusing on these regions. Further, most key terrain in these regions lies within easy distance of seaports, Afghanistan being the exception. The Afghanistan campaign (Operation Enduring Freedom) is unusual in that all forces and most supplies had to be transported by air. Even if Pakistan had granted unrestricted use of its territory, a poorly developed infrastructure would still have compelled the United States to rely primarily on airlift.

Chapter Five

CONCLUSIONS

Army transformation efforts seek to turn the Army (first the Stryker Brigade Combat Team, then the Objective Force) into a force that can deploy globally in 96 hours. However, the analysis in this report suggests that a force with over 1,000 vehicles cannot be deployed by air from CONUS to the far reaches of the globe in four days. With some mobility enhancements, it will be possible to achieve deployment timelines on the order of one to two weeks, which is quite rapid for a motorized force. Specifically, this analysis found that the combination of CONUS bases (particularly Fort Polk), an SBCT forward-based in Germany, and regional preposition sites in Guam and Diego Garcia offers the ability to deploy the SBCT by air or sea to key regions in 5 to 14 days. Figure 5.1 illustrates this for scenarios in East Asia, South America, Africa, Europe, and Southwest Asia.

KEY FINDINGS

- Large U.S. joint operations have historically been concentrated in relatively few regions: Europe, Latin America, the Persian Gulf, and Asia.
- The global war on terrorism is a "wild card" that could lead to operations in more-remote locales than the historical concentration. For most such operations, the forces involved are likely to be smaller than the SBCT.

RANDMR1606-5.1

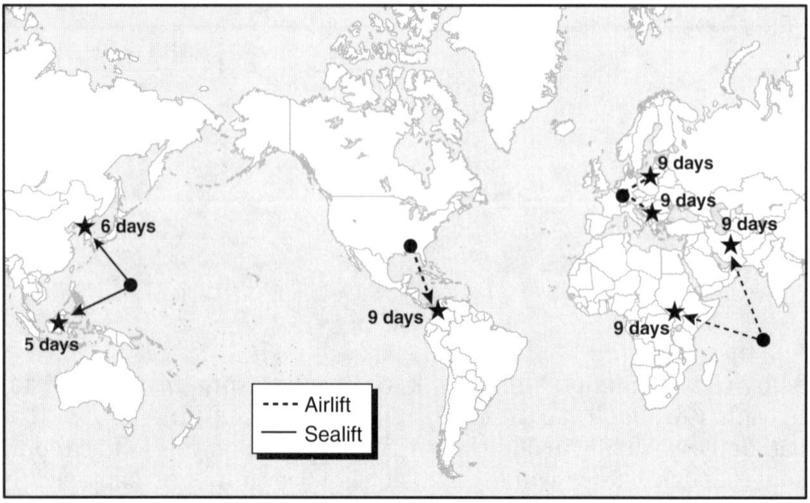

Figure 5.1—Regional Basing, Showing Rapid Deployment Times to Key Areas

- Security challenges have typically developed over a time frame of months or years, allowing for prepositioning and other regional defensive measures that reduce the need for rapid deployment from CONUS.
- Prepositioning of equipment or overseas basing of forces is the single most effective way to increase the strategic responsiveness of U.S. Army forces for operations in key regions. From these preposition sites, the choice of airlift or sealift will depend on the scenario.
- In general, deep interior deployments favor airlift; littoral scenarios favor sealift. In some cases, neither an airfield nor a port will be particularly close to the area of operations, and long road marches either way will be required. In such cases, a detailed analysis of road networks and other local considerations would be necessary to determine the preferred deployment mode.
- For littoral deployments from preposition sites to unsophisticated ports, fast, shallow-draft ships like the catamaran ferries currently being tested by DoD appear to offer the fastest and

most robust option. However, their shorter range may require more preposition sites than larger ships or new at-sea refueling concepts. The ability of shallow-draft ships to use smaller ports averts the time delays, complications, and weather constraints associated with lighterage.

RECOMMENDATIONS FOR THE USAF

Army efforts to develop medium-weight forces offer Joint Task Force or theater commanders capabilities not resident in current light or heavy forces and should be supported by the USAF. Although the more ambitious air deployment objectives may not be feasible, air transport remains the fastest option for some contingencies. Fast sealift is promising for littoral operations. Even then, airlift is likely to be called upon to move critical personnel and equipment. For example, Army port operations and security personnel and equipment might move by air to prepare a port for the arrival of the SBCT. Special forces might move by air to conduct supporting reconnaissance, direct action, or other special missions. In forced-entry scenarios, airborne forces might seize a port for the SBCT. Finally, airlift is likely to play a critical role in high-priority resupply and support operations.

More broadly, we note that the Air Force has a stake in Army transformation efforts. The Army envisions future forces operating in ways that are likely to require closer air-ground cooperation on intelligence, surveillance, and reconnaissance (ISR); lift; and precision fires. We recommend that Air Force and Army leaders initiate a dialogue on these issues of mutual concern. The Army would greatly benefit from the expertise the USAF brings on air deployment, ISR, survivability of transport aircraft, and air-to-ground fires.

The USAF is beginning to develop new concepts for air-to-ground operations and would benefit greatly from Army expertise on land operations and from the substantial effort the Army has already invested in developing new concepts for the future battlefield. Airmen and soldiers working together may also develop new concepts for air mobility that overcome some of the constraints discussed in this report. Land-warfare theorists are developing concepts that are ambitious. Airmen can help the Army by identifying some concepts that may never be operationally feasible; at the same time, the Army creative efforts may spur airmen to develop complementary concepts and capabilities that they would not otherwise have pursued.

Appendix A

DEPTH OF U.S. OPERATIONS

In Chapter Two, we noted that airlift is generally preferred for operations hundreds of miles from ports, and that sealift is most attractive for littoral contingencies. To provide some historical context, we studied several past operations and measured the greater distance U.S. forces operated from resupply ports.

In past operations (excluding the Civil War), U.S. joint forces accomplished their missions at depths from the littorals that normally did not exceed 600–700 kilometers (km). During World War II, in early September 1944, the Allied line of communications ran from Cherbourg to eastern France near the German border, a depth of about 620 km.[1] During the Korean War, in September 1950, U.S. forces advanced from the Pusan perimeter to the vicinity of Suwon, a depth of about 310 km.[2] During a largely unopposed advance to the Yalu River, U.S. X Corps advanced to a depth of about 270 km.[3]

[1]Following the invasion of Normandy on June 6, a tenacious German resistance frustrated the Allies until they closed the Falaise gap on August 22. Thereafter, Allied forces advanced against little resistance until early September, when they came to the end of their logistical tether. The road distance for Cherbourg-Paris-Metz is about 620 km. British forces entered Antwerp on September 4, and it rapidly became a major port of entry for Allied forces, greatly shortening the line of communications.

[2]On September 15, U.S. forces landed at Inchon in Operation Chromite, leading to a rapid collapse of North Korean resistance. On September 26, elements of X Corps (1st Marine Division and 7th Infantry Division) linked with elements of the Eighth Army in the vicinity of Osan-Suwon. The road distance from Pusan to Suwon is approximately 310 km.

[3]On October 20, elements of X Corps began landing at Wonsan on the Sea of Japan. On November 21, the 17th Infantry occupied Hyesanjin on the Yalu River. The road distance from Wonsan to Hyesanjin is approximately 270 km.

During Operation Desert Shield/Desert Storm (August 1990–February 1991), the 1st Armored Division operated to a depth of approximately 720 km and the 101st Airborne Division (Air Assault) operated to a depth of about 850 km.[4]

As in the past, most objectives will probably lie within operational reach of littorals, but there may be important exceptions. Had the U.S. decided, in April 1994, to stop genocide in landlocked Rwanda, it would have airlifted forces, rather than conducting a 1,500-km road march.[5] Had the U.S. decided to stop the Bosnian Serb assault on the "safe area" of Srebrenica during July 1995, it would have airlifted its forces, presumably using helicopters for the last increment.[6] Advance from the Adriatic coast would have been wholly impractical because the road wound through difficult terrain controlled by Bosnian Serb forces to a depth of about 320 km.

Similarly, had the U.S. decided to initiate a land offensive into Kosovo during fall 1999, the distances would have been fairly short, but the United States would still have relied heavily on airlift to ensure success before winter. The route from Durrës over Kukes to Pristina is only about 260 km, not accounting for numerous switchbacks, but it traverses very rugged terrain on the Albanian-Kosovo border and includes numerous bridges of doubtful capacity. The route over Thessaloniki and Skopje is only a little longer and much easier, but it was expected to serve primarily British forces who would be advancing up that axis.

If operations in landlocked Afghanistan in fall 2001 had required large U.S. ground forces, the United States would probably have

---

[4]The 1st Armored Division advanced from the coast, notionally from the major port of Ad Dammam, to positions in the An Nafud Desert, and then to positions in northern Kuwait astride the north-south highway. The distance for Ad Dammam–An Nafud–Kuwait is approximately 720 km. The 101st Airborne Division (Air Assault) advanced farther into An Nafud, then north to positions in the vicinity of the Euphrates River. The distance for Ad Dammam–An Nafud–Euphrates is approximately 850 km.

[5]The best land line of communications is probably Mombasa-Nairobi-Kampala-Kigali, implying a road distance of at least 1,500 km.

[6]See United Nations, *Report of the Secretary-General Pursuant to General Assembly Resolution 53/35, The Fall of Srebrenica,* New York: A/54/549, November 15, 1999.

made heavy use of airlift, considering the difficult alternatives.[7] Even if Pakistan, Iran, or Russia/Uzbekistan had allowed transit, the overland options were not attractive. As Figure A.1 illustrates, the shortest route (Karachi-Quetta-Kandahar-Kabul) would have required movement over 1,400 km, much of it on poor roads.[8] The Iranian option is some 1,700 km to Kabul.[9] From the north, the most direct route would have used the Black Sea port of Novorossiysk, then rail movement to Volgograd, around the Caspian Sea, and down through Uzbekistan—a trek of some 3,750 km.[10] Sustaining land lines of communication of these lengths would have been very difficult and time-consuming. It is unlikely that the United States could have moved significant forces via these routes before spring 2002. But in the event, the U.S. required only modest-sized land forces to ensure the victory of anti-Taliban forces and to hunt for al Qaeda members.

Table A.1 summarizes depth in selected operations. Operations in Korea and the Persian Gulf actually occurred. Rwanda, Bosnia, and Kosovo are hypothetical examples of operations that the United States might have conducted had a different decision been made.

---

[7]It is worth noting that the U.S. Marine Brigade Task Force that deployed to the region south of Kandahar moved, and was sustained entirely, by air. If the U.S. Army had had a forward-deployed SBCT in October 2001, this would have been a good mission for it. As it was, the two Marine Expeditionary Units afloat off the coast of Pakistan were the only forces that possessed the combination of rapid deployment, significant firepower, and tactical mobility that Central Command (CENTCOM) sought.

[8]The two major routes are Karachi-Sukkur-Sibi-Quetta-Kandahar-Krasny-Kabul (1,400 km) and Karachi–Dera Ghazi Khan–Peshwar-Jalalabad-Kabul (1,600 km). The first route leads directly to Kandahar, the most important Taliban stronghold. The second route leads past Islamabad, where pro-Taliban elements would presumably protest U.S. deployment, and through the Khyber Pass, an extremely difficult stretch of road.

[9]The route for Bandar Beheshti–Zahedan-Zabol-Farah-Kandahar-Kabul is about 1,700 km. Moreover, the port of Bandar Beheshti is poorly developed.

[10]The route for Novorossiysk-Krasnodar-Rostov-Volgograd-Astrakhan-Bukhara-Termez–Mazar-e Sharif–Kabul is at least 3,750 km. Novorossiysk is a well-developed port on the Black Sea.

RANDMR1606-A.1

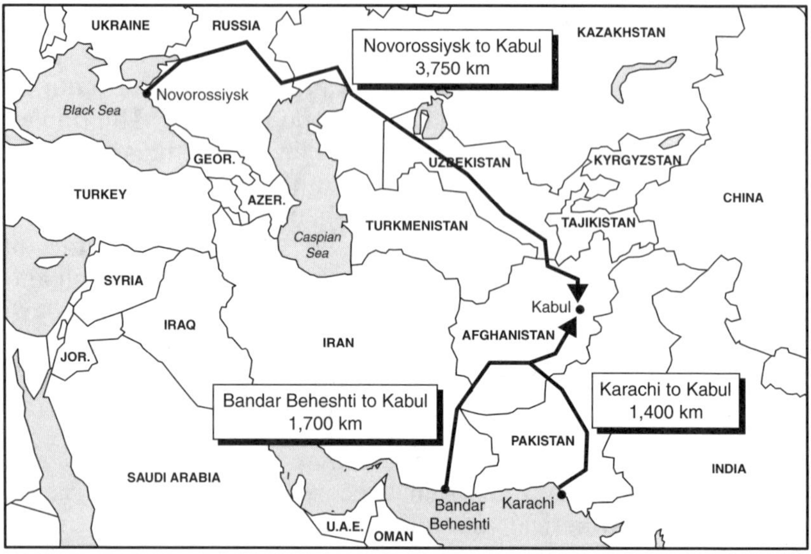

Figure A.1—Routes into Afghanistan

Table A.1

Depth in Selected Operations

		Distance	
Advance to:	Route	km	nmi
Seoul, Korea 1950	Pusan-Taejon-Suwon	310	168
Hyesanjin, Korea 1950	Wonsan-Hamhung-Hyesanjin	270	145
As Samawah, Iraq 1991	Ad Dammam–An Nafud–As Samawah	850	460
Northern Kuwait 1991	Ad Dammam–An Nafud–northern Kuwait	720	390
Kigali, Rwanda 1994	Mombasa-Nairobi-Kampala-Kigali	1,500	810
Srebrenica, Bosnia 1995	Ploce-Mostar-Sarajevo-Srebrenica	320	175
Pristina, Kosovo 1999	Durrës-Kukes-Prizren-Pristina	260	140
Pristina, Kosovo 1999	Thessaloniki-Skopje-Pristina	295	160

Appendix B

COMPONENTS OF DEPLOYMENT TIMES FOR ALL SCENARIOS FROM CHAPTER TWO

This appendix provides additional details for the deployment scenarios discussed in Chapter Two. Table B.1 lists aerial ports of embarkation (APOEs), in-transit bases, distances of each leg, aerial ports of debarkation (APODs), and total deployment times for airlift from the continental United States (CONUS) and forward bases. Table B.2 lists sea ports of embarkation (SPOEs); final destinations; deployment distances; ship-loading, transit, and unloading times; assembly and road-march times; and total deployment times.

Table B.1

Components of Airlift Deployment Times for All Scenarios in Figure 2.2

APOE	En Route Stops	Distance (nmi)	APOD	Deployment Time
Hickam AFB, HI	Andersen AFB, Guam	3,302 + 1,791 = 5,093	Tanjung Bara, Indonesia	14.0 days
Barksdale AFB, LA	None	2,006	Bogota, Colombia	8.8 days; MOG limited
Barksdale AFB, LA	Lima, Peru	2,847 + 2,368 = 5,215	Puenta Arenas, Chile	14.3 days
Barksdale AFB, LA	Lajes Field, Azores and Moron AFB, Spain	3,214 + 1,021 + 2,770 = 7,005	Riyadh, Saudia Arabia	19.6 days
Barksdale AFB, LA	Lajes Field, Azores and Aviano, Italy	3,214 + 1,814 + 2,615 = 7,643	Kandahar, Afghanistan	21.1 days
Barksdale AFB, LA	Porto Velho, Brazil and Rio de Janeiro, Brazil	3,010 + 1,454 + 3,290 = 7,754	Cape Town, South Africa	21.4 days
Barksdale AFB, LA	Keflavic, Iceland	3,180 + 1,408 = 4,588	Riga, Latvia	12.8 days
McChord AFB, WA	Elmendorf AFB, AK	1,291 + 3,295 = 4,586	Osan, Korea	12.8 days
Barksdale AFB, LA	Lajes Field, Azores	3,214 + 2,223 = 5,437	Skopje, Macedonia	14.9 days
Barksdale AFB, LA	Lajes Field, Azores and Cairo, Egypt	3,214 + 2,901 + 1,932 = 8,047	Kigali, Rwanda	22.1 days
Guam	None	1,791	Tanjung Bara, Indonesia	8.8 days; MOG limited
Diego Garcia	None	2,443	Riyadh	8.8 days; MOG limited
Diego Garcia	None	2,374	Kandahar	8.8 days; MOG limited
Diego Garcia	None	3,370	Cape Town, South Africa	9.1 days; MOG limited
Ramstein AB, Germany	None	739	Riga, Latvia	8.7 days; MOG limited
Guam	None	1,705	Osan, Korea	8.7 days; MOG limited
Ramstein AB, Germany	None	738	Skopje, Macedonia	8.7 days; MOG limited
Diego Garcia	None	2,550	Kigali, Rwanda	8.8 days; MOG limited

RAND*MR1606-TB.1*

Table B.2

Components of Sealift Deployment Times for All Scenarios in Figure 2.6

SPOE	SPOD	Final Destination	Distance (nmi)	Load ships (days)	Steam (days)	Canal Delay (days)	Unload (days)	Assemble (days)	Road March (days)	Deployment Time (days)
Honolulu, HI	Tanjung Bara, Indonesia	Tanjung Bara, Indonesia	5,089	1.7	7.9	NA	3.1	1	NA	13.7
Beaumont, TX	Buenaventura, Colombia	Bogota, Colombia	1,796	1.7	1.9	1	3.1	1	1.9	10.6
Beaumont, TX	Santa Marta, Colombia	Bogota, Colombia	1,562	1.7	1.6	NA	1.7	1	5.3	11.3
Honolulu, HI	Punta Arenas, Chile	Punta Arenas, Chile	6,363	1.7	9.8	NA	3.1	1	NA	15.6
Beaumont, TX	Punta Arenas, Chile	Punta Arenas, Chile	5,689	1.7	8.8	1	3.1	1	NA	15.6
Beaumont, TX	Ad Dammam, Saudi Arabia	Ad Dammam, Saudi Arabia	9,879	1.7	15.2	1	1.7	1	NA	20.6
Beaumont, TX	Karachi, Pakistan	Kandahar, Afghanistan	9,517	1.7	14.7	1	1.7	1	7	27.1
Beaumont, TX	Cape Town, South Africa	Cape Town, South Africa	7,640	1.7	11.8	NA	1.7	1	NA	16.2
Beaumont, TX	Riga, Latvia	Riga, Latvia	5,717	1.7	8.8	NA	1.7	1	NA	13.2
Seattle, WA	Pusan, Korea	Seoul, Korea	4,689	1.7	7.2	NA	1.7	1	1.3	12.9
Beaumont, TX	Thessaloniki, Greece	Skopje, Macedonia	6,378	1.7	9.8	NA	1.7	1	0.9	15.1
Beaumont, TX	Mombasa, Kenya	Kigali, Rwanda	9,684	1.7	14.9	1	3.1	1	7.2	28.9
Guam	Tanjung Bara, Indonesia	Tanjung Bara, Indonesia	1,790	Maritime prepo	1.9	NA	1.7	1	NA	4.6
Diego Garcia	Ad Dammam, Saudi Arabia	Ad Dammam, Saudi Arabia	2,619	Maritime prepo	2.7	NA	1.7	1	NA	5.4
Diego Garcia	Karachi, Pakistan	Kabul, Afghanistan	1,960	Maritime prepo	2.0	NA	1.7	1	7	11.7
Diego Garcia	Cape Town, South Africa	Cape Town, South Africa	3,448	Maritime prepo	3.6	NA	1.7	1	NA	6.3
Bremerhaven, Germany	Riga, Latvia	Riga, Latvia	934	Maritime prepo	1.0	NA	1.7	1	NA	3.7
Guam	Pusan, Korea	Seoul, Korea	1,627	Maritime prepo	1.7	NA	1.7	1	1.3	5.7
Livorno	Thessaloniki, Greece	Skopje, Macedonia	1,224	Maritime prepo	1.3	NA	1.7	1	0.9	4.9
Diego Garcia	Mombasa, Kenya	Kigali, Rwanda	1,961	Maritime prepo	2.0	NA	1.7	1	7.2	11.9

RANDMR1606-TB.2

BIBLIOGRAPHY

Agence France-Presse, "Gun-Toting U.S. Soldiers Have Apologized, Says Macapagal," *Manila Times,* February 2002.

Asmus, Ronald D., *The New U.S. Strategic Debate,* Santa Monica, Calif.: RAND, MR-240-1-A, 1994.

BBC News, "Guide to Philippines Conflict," December 6, 2001, available at http://new.bbc.co.uk/hi/english/world/asia-pacific/newsid_169500/1695576.stm.

Binyon, Michael, "How NATO Members Are Facing the Flak," *The Times* (London), March 27, 1999.

Blank, Stephen J., *U.S. Military Engagement with Transcaucasia and Central Asia,* Carlisle, Penn.: U.S. Army War College, Strategic Studies Institute, 2000.

Blechman, Barry M., and Stephen S. Kaplan, *Force Without War: U.S. Armed Forces as a Political Instrument,* Washington, D.C.: The Brookings Institution, 1978.

Bonner, Raymond, "U.S. and Philippine Governments Revive Old Relationship," *The New York Times,* March 4, 2002a.

Bonner, Raymond, "'Yankee Don't Go' Is Message in Philippines Antiterror Rally," *The New York Times,* February 25, 2002b.

Bostock, Ian, "USMC Charters High-Speed Catamaran," *Jane's Defence Weekly,* 11 July 2001, p. 3.

Bowden, Mark, *Black Hawk Down,* New York: Atlantic Monthly Press, 1999.

Burk, James, "Public Support for Peacekeeping in Lebanon and Somalia: Assessing the Casualties Hypothesis," *Political Science Quarterly,* Vol. 114, No. 1, 1999.

Bush, George W., "Statement by the President in His Address to the Nation," national television, 8:30 p.m. EDT, available at http://www.whitehouse.gov/news/releases/2001/09/20010911-16.html.

Byman, Daniel, Shahram Chubin, Anoushiravan Ehteshami, and Jerrold D. Green, *Iran's Security Policy in the Post-Revolutionary Era,* Santa Monica, Calif.: RAND, MR-1320-OSD, 2001.

Byman, Daniel L., and Jerrold D. Green, *Political Violence and Stability in the States of the Northern Persian Gulf,* Santa Monica, Calif.: RAND, MR-1021-OSD, 1999.

Byman, Daniel L., and Matthew C. Waxman, *Confronting Iraq: U.S. Policy and the Use of Force Since the Gulf War,* Santa Monica, Calif.: RAND, MR-1146-OSD, 2000.

Carlucci, Frank, Robert Hunter, and Zalmay Khalilzad, eds., *Taking Charge: A Bipartisan Report to the President-Elect on Foreign Policy and National Security,* Santa Monica, Calif.: RAND, MR-1306-RC, 2001.

Chairman, Joint Chiefs of Staff (CJCS), *Joint Vision 2020,* Washington, D.C.: J-5, Strategic Plans and Policy, June 2000.

Chalk, Peter, *Australian Foreign and Defense Policy in the Wake of the 1999/2000 East Timor Intervention,* Santa Monica, Calif.: RAND, MR-1409-SRF, 2001.

Chen, Shuxun, and Charles Wolf, Jr., eds., *China, the United States, and the Global Economy,* Santa Monica, Calif.: RAND, MR-1300-RC, 2001.

Clark, Wesley K., *Waging Modern War: Bosnia, Kosovo, and the Future of Combat,* New York: Public Affairs, 2001.

Clinton, William J., *Executive Order 13129 of July 4, 1999, Blocking Property and Prohibiting Transactions with the Taliban*, Washington, D.C.: Presidential Documents, *Federal Registry*, Vol. 64, No. 129, July 7, 1999.

Conception, Faber, "Akbar Gets U.S. Assurance on Projects," *Manila Times*, April 30, 2002.

Cordesman, Anthony H., *Bahrain, Oman, Qatar, and the UAE: Challenges of Security*, Boulder, Colo: Westview Press, 1997.

Farah, Douglas, "Digging Up Congo's Dirty Gems," *The Washington Post*, December 30, 2001, pp. A1, A16.

Feil, Scott R., *Preventing Genocide: How the Early Use of Force Might Have Succeeded in Rwanda*, New York: Carnegie Commission, 1998.

Fitchett, Joseph, interview with Greek Foreign Minister George Papandreou, contained in "A Call for Bolder Vision in Balkans," *The International Herald Tribune*, May 22, 1999, p. 5.

Flaherty, Mary Pat, David B. Ottaway, and James V. Grimaldi, "How Afghanistan Went Unlisted as Terrorist Sponsor," *The Washington Post*, November 5, 2001, pp. A1, A15.

Fuller, Graham E., *"The Center of the Universe": The Geopolitics of Iran*, Boulder, Colo.: Westview Press, 1991.

Fuller, Graham E., and Ian O. Lesser, *A Sense of Siege: The Geopolitics of Islam and the West*, Boulder, Colo.: Westview Press, 1995.

Fuller, Graham E., and Bruce R. Pirnie, *Iran: Destabilizing Potential in the Persian Gulf*, Santa Monica, Calif.: RAND, MR-793-OSD, 1996.

Gong, Gerrit W., ed., *Taiwan Strait Dilemmas: China-Taiwan-U.S. Policies in the New Century*, Washington, D.C.: Center for Strategic and International Studies, 2000.

Gordon, John, et al., "Kosovo Revisited: Responding to the Future Threat," Santa Monica, Calif.: unpublished RAND research.

Gourevitch, Philip, *We Wish to Inform You That Tomorrow We Will Be Killed with Our Families: Stories from Rwanda,* New York: Farrar, Straus and Giroux, 1998.

Gritton, Eugene C., Paul K. Davis, Randall Steeb, and John Matsumura, *Ground Forces for a Rapidly Employable Joint Task Force: First-Week Capabilities for Short-Warning Conflicts,* Santa Monica, Calif.: RAND, MR-1152-OSD/A, 2000.

Haughney, Christine, "Unearthing an Elusive Death Count," *The Washington Post,* October 24, 2001, p. A4.

Headquarters, Department of the Army, *Armor and Mechanized-Based Opposing Force Operational Art,* Washington, D.C.: Field Manual 100-61, 26 January 1998.

Headquarters, Department of the Army, *The Army,* Washington, D.C.: Field Manual 1, June 14, 2001a.

Headquarters, Department of the Army, "Army Transformation, Interim Force," briefing prepared by MG Jim Grazioplene, USA, Washington, D.C., October 17, 2000a.

Headquarters, Department of the Army, "Army Transformation, Objective Force," briefing prepared by LTG Paul J. Kern, USA, Washington, D.C., October 17, 2000b.

Headquarters, Department of the Army, *The Army Vision: Soldiers on Point for the Nation, Persuasive in Peace, Invincible in War,* Washington, D.C., August 30, 2001b, available at http://www.army.mil/vision.

Headquarters, Department of the Army, *Concept for the Objective Force,* Washington, D.C.: White Paper with foreword by General Eric K. Shinseki, Chief of Staff, November 2001c.

Headquarters, Department of the Army, *Operational Requirements Document for a Family of Interim Armored Vehicles (IAV), Prepared for Milestone I Decision,* Washington, D.C., April 6, 2000c.

Headquarters, Department of the Army, *Operations,* Washington, D.C.: Field Manual 3-0, June 14, 2001d.

Headquarters, Department of the Army, "Strategic Responsiveness Study, A Holistic View," briefing prepared by BG Charlie Fletcher, Director of Force Projection & Distribution, Washington, D.C.: Office of the Deputy Chief of Staff for Logistics, 2001e.

Headquarters, Department of the Army, *Transportation Reference Data,* Washington, D.C.: Field Manual 55-15, 27 October 1997.

Headquarters, U.S. Air Force (HQ USAF), Office of the Deputy Chief of Staff for Air and Space Operations, "AEF Sustainment Estimates," unpublished briefing, September 26, 2000.

Headquarters, U.S. Air Force, *Air Mobility Planning Factors,* Washington, D.C.: Air Force Pamphlet 10-1403, 1 March 1998, available at http://afpubs.hq.af.mil/pubfiles/af/10/afpam10-1403/afpam10-1403.pdf.

Headquarters, U.S. Air Force Europe/AMOCC, "Allied Force: Intra-Theater Air Mobility," briefing prepared by MAJ Martin Beard, USA, July 1999.

Hunter, Shireen T., *Central Asia Since Independence,* Westport, Conn: Praeger, 1996.

International Institute for Strategic Studies, *Strategic Survey 1999/2000,* London, 2000.

Joint Chiefs of Staff, *Joint Tactics, Techniques, and Procedures for Joint Logistics Over-the-Shore (JLOTS),* Washington, D.C.: Joint Publication 4-01.6, 12 November 1998.

Kaplan, Stephen S., *Force Without War: U.S. Armed Forces as a Political Instrument,* Washington, D.C.: The Brookings Institution, 1978.

Kelso, Admiral Frank B. II, Chief of Naval Operations, and Commandant of the Marine Corps General C. E. Mundy, Jr., Department of the Navy, *. . . From the Sea: Preparing the Naval Service for the 21st Century,* Washington, D.C.: U.S. Navy and U.S. Marine Corps, 1992.

Kennedy, John F., Address to the Nation, broadcast on October 22, 1962.

Khalilzad, Zalmay, and Ian O. Lesser, eds., *Sources of Conflict in the 21st Century: Regional Futures and U.S. Strategy*, Santa Monica, Calif.: RAND, MR-897-AF, 1998.

Khalilzad, Zalmay, David T. Orletsky, Jonathan D. Pollack, Kevin Pollpeter, Angel Rabasa, David A. Shlapak, Abram N. Shulsky, and Ashley J. Tellis, *The United States and Asia: Toward a New U.S. Strategy and Force Posture*, Santa Monica, Calif.: RAND, MR-1315-AF, 2001.

Khalilzad, Zalmay, Abram N. Shulsky, Daniel Byman, Roger Cliff, David T. Orletksy, David A. Shlapak, and Ashley J. Tellis, *The United States and a Rising China: Strategic and Military Implications*, Santa Monica, Calif.: RAND, MR-1082-AF, 1999.

Kissinger, Henry, *Years of Upheaval,* Boston, Mass.: Little, Brown and Company, 1982.

Kozaryn, Linda D., "Attacks Continue in Afghanistan; Philippines Training Under Way," *Armed Forces Information Service,* April 24, 2002.

Kugler, Richard L., *Changes Ahead: Future Directions for the U.S. Overseas Military Presence*, Santa Monica, Calif.: RAND, MR-956-AF, 1998.

Kuperman, Alan, *The Limits of Humanitarian Intervention,* Washington, D.C.: The Brookings Institution, 2001.

LaFraniere, Sharon, "U.S. to Train 1,200 Georgian Soldiers in Terror Battle," *The Washington Post,* March 2, 2002, pp. A14, A20.

Larrabee, F. Stephen, Jerrold D. Green, Ian O. Lesser, and Michele Zanini, *NATO's Mediterranean Initiative: Policy Issues and Dilemmas*, Santa Monica, Calif.: RAND, MR-957-IMD, 1998.

Larson, Eric V., *Casualties and Consensus: The Historical Role of Casualties in Domestic Support of U.S. Military Operations*, Santa Monica, Calif.: RAND, MR-726-RC, 1996.

Lehman, John F., Jr., *Command of the Seas*, New York: Charles Scribner's Sons, 1988.

Lesser, Ian O., *NATO Looks South: New Challenges and New Strategies in the Mediterranean*, Santa Monica, Calif.: RAND, MR-1126-AF, 2000.

Lesser, Ian O., Jerrold D. Green, F. Stephen Larrabee, and Michele Zanini, *The Future of NATO's Mediterranean Initiative: Evolution and Next Steps*, Santa Monica, Calif.: RAND, MR-1164-SMD, 2000.

Lipton, Eric, "A New Count of the Dead, but Little Sense of Relief," *The New York Times*, December 2, 2001, available at http://college4.nytimes.com/guests/articles/2001/12/02/887970.xml.

Lloyd's of London, *Ports of the World*, London, 1994, 2001.

Loeb, Vernon, and Peter Slevin, "U.S. Special Forces to Advise Yemeni Military," *The Washington Post*, March 2, 2002, p. A20.

Lund, John, Ruth Berg, and Corinne Replogle, *Project AIR FORCE Analysis of the Air War in the Gulf: An Assessment of Strategic Airlift Operational Efficiency*, Santa Monica, Calif.: RAND, R-4269/4-AF, 1993.

Matsumura, John, Randall Steeb, John Gordon, Russell W. Glenn, Thomas J. Herbert, and Paul Steinberg, *Lightning Over Water: Sharpening America's Light Forces for Rapid-Reaction Missions*, Santa Monica, Calif.: RAND, MR-1196-A/OSD, 2000.

Metz, Steven, *Refining American Security in Africa*, Carlisle, Penn.: U.S. Army War College, Strategic Studies Institute, 2000.

Mueller, John E., *Policy and Opinion in the Gulf War*, Chicago: University of Chicago Press, 1994.

Mueller, John E., *War, Presidents, and Public Opinion*, New York: John Wiley & Sons, 1973.

Nardulli, Bruce, Walt L. Perry, Bruce Pirnie, John Gordon, and John G. McGinn, *Disjointed War: Military Operations in Kosovo, 1999*, Santa Monica, Calif.: RAND, MR-1406-A, 2002.

National Defense University, *Strategic Assessment 1999: Priorities for a Turbulent World*, Washington, D.C.: Institute for National Strategic Studies, 1999.

National Imagery and Mapping Agency (NIMA), *World Port Index,* Pub 150, http://164.214.12.145/pubs/pubs_j_wpi_sections.html, November 1, 2002.

Naval Postgraduate School, "Terrorist Group Profiles," Monterey, Calif.: Dudley Knox Library, available at http://web.nps.navy.mil/~library/tgp/asc.htm.

Norton, Daniel M., "IBCT Deployment: An Initial Assessment," Santa Monica, Calif.: unpublished RAND research.

Oliker, Olga, *Russia's Chechen Wars 1994–2000: Lessons from Urban Combat,* Santa Monica, Calif.: RAND, MR-1289-A, 2001.

Pan, Philip P., "Some Filipinos Say U.S. Presence May Fuel Rebel Support," *The Washington Post,* February 7, 2002, p. 13.

Peach, Stuart, ed., *Perspectives on Air Power: Air Power in Its Wider Context,* London: The Stationery Office, 1998.

Perlez, Jane, "Philippine Troops Eagerly Await U.S. Help and Arms," *The New York Times,* February 12, 2002, p. 1.

Peters, F. Whitten, Secretary of the Air Force, and Chief of Staff of the Air Force General Michael E. Ryan, *America's Air Force Vision 2020: Global Vigilance, Reach, and Power,* Washington, D.C.: U.S. Air Force, 2001, available at http://www.af.mil/vision.

Pollack, Jonathan D., and Chung Min Lee, *Preparing for Korean Unification: Scenarios and Implications,* Santa Monica, Calif.: RAND, MR-1040-A, 1999.

Polson, William, "Navy Goes Down Under, Explores Future of Amphibious Warfare," available at http://www.c7f.navy.milnews/2000/09/16/html, posted September 16, 2000; website last accessed June 24, 2002.

Powell, Colin, with Joseph E. Persico, *My American Journey: An Autobiography,* New York: Random House, 1995.

Power, Samantha, "Bystanders to Genocide: Why the United States Let the Rwandan Tragedy Happen," *The Atlantic Monthly,* Vol. 288, No. 2, September 2001, pp. 84–108.

Rabasa, Angel, and Peter Chalk, *Colombian Labyrinth: The Synergy of Drugs and Insurgency and Its Implications for Regional Stability,* Santa Monica, Calif.: RAND, MR-1339-AF, 2001a.

Rabasa, Angel, and Peter Chalk, *Indonesia's Transformation and the Stability of Southeast Asia,* Santa Monica, Calif.: RAND, MR-1344-AF, 2001b.

Ramey, Timothy L., *Lean Logistics: High-Velocity Logistics Infrastructure and the C-5 Galaxy,* Santa Monica, Calif.: RAND, MR-581-AF, 1999.

Ricks, Thomas E., "Allies Step Up Somalia Watch, U.S. Aims to Keep al Qaeda at Bay," *The Washington Post,* January 4, 2002a, p. A1.

Ricks, Thomas E., "U.S. Eyes Military Assistance for Yemen," *The Washington Post,* February 28, 2002b, p. 1.

Rumsfeld, Donald, *Quadrennial Defense Review Report,* Washington, D.C.: U.S. Department of Defense, 30 September 2001.

Schelling, Thomas C., *Arms and Influence,* New Haven, Conn.: Yale University Press, 1966.

Schwarzkopf, H. Norman, with Peter Petre, *General H. Norman Schwarzkopf: The Autobiography: It Doesn't Take a Hero,* New York: Bantam Books, 1993.

Scobell, Andrew, and Larry M. Wortzel, *The Asia Pacific in the U.S. National Security Calculus for a New Millennium,* Carlisle, Penn.: U.S. Army War College, Strategic Studies Institute, 2000.

Shinseki, General Eric K., USA, "The Army Transformation: A Historic Opportunity," *Army,* October 2000, pp. 21–30.

Shinseki, General Eric K., "The Army Vision: A Status Report," *Army,* October 2001.

Shlapak, David A., David T. Orletsky, and Barry A. Wilson, *Dire Strait? Military Aspects of the China-Taiwan Confrontation and Options for U.S. Policy,* Santa Monica, Calif.: RAND, MR-1217-SRF, 2000.

Sokolsky, Richard, and Tanya Charlick-Paley, *NATO and Caspian Security: A Mission Too Far?* Santa Monica, Calif.: RAND, 1074-AF, 1999.

Sokolsky, Richard, Stuart Johnson, and F. Stephen Larrabee, *Persian Gulf Security: Improving Allied Military Contributions,* Santa Monica, Calif.: RAND, MR-1245-AF, 2000.

Sokolsky, Richard, Angel Rabasa, and C. R. Neu, *The Role of Southeast Asia in U.S. Strategy Toward China,* Santa Monica, Calif.: RAND, MR-1170-AF, 2000.

Solomon, Gerald B., *The NATO Enlargement Debate, 1990–1997: The Blessings of Liberty,* Westport, Conn.: Praeger, 1998.

Stucker, James P., and Ruth T. Berg, *Understanding Airfield Capacity for Airlift Operations,* Santa Monica, Calif.: RAND, MR-700-AF/OSD, 1998.

Swaine, Michael D., *Taiwan's National Security, Defense Policy, and Weapons Procurement Process,* Santa Monica, Calif.: RAND, MR-1128-OSD, 1999.

Swaine, Michael D., and Ashley J. Tellis, *Interpreting China's Grand Strategy,* Santa Monica, Calif.: RAND, MR-1121-AF, 2000.

Szayna, Thomas S., *NATO Enlargement 2000–2015: Determinants and Implications for Defense Planning and Shaping,* Santa Monica, Calif.: RAND, MR-1243-AF, 2001.

Tellis, Ashley J., *India's Emerging Nuclear Posture: Between Recessed Deterrent and Ready Arsenal,* Santa Monica, Calif.: RAND, MR-1127-AF, 2001.

Thatcher, Margaret, *The Downing Street Years,* New York: HarperCollins, 1993.

"The Times Poll: Americans Back Bush Decision Overwhelmingly," *Los Angeles Times,* January 19, 1991, p. A1.

Truman, Harry S., "Statement by the President in His Address to the Nation," national television, 8:30 p.m. EDT, available at http://www.whitehouse.gov/news/releases/2001/09/20010911–16.html.

United Nations, *Report of the Secretary-General Pursuant to General Assembly Resolution 53/35, The Fall of Srebrenica,* New York: A/54/549, November 15, 1999.

U.S. Department of Defense, *Annual Report to the President and the Congress 2001,* Washington, D.C., 2001, available at http://www.dtic.mil/execscc/adr2001.

U.S. Department of State, *2001 Report on Foreign Terrorist Organizations,* Washington, D.C.: Office of the Coordinator for Counterterrorism, October 5, 2001, available at http://www.state.gov/s/ct/ris/rpt/fto.

U.S. General Accounting Office, *Airlift Operations of the Military Airlift Command During the 1973 Middle East War,* Washington, D.C., 1975.

U.S. Joint Forces Command, *A Concept for Rapid Decisive Operations,* Norfolk, Va.: J-9, Joint Futures Laboratory, RDO White Paper Version 2.0, October 25, 2001.

U.S. Joint Forces Command, *Strategic Deployment: A White Paper for Joint Experimentation,* Norfolk, Va.: J-9, Concepts Division, Version 1.0, May 10, 2000.

U.S. Marine Corps, *MEF Planner's Reference Manual,* Quantico, Va.: MAGTAF Staff Training Program, MSTP Pamphlet 5-0.3, 1999.

U.S. Transportation Command, *Interim Brigade Combat Team Air Mobility Deployment Analysis,* Scott AFB, Ill., April 2002.

Wang, Hui, "U.S.–China Bonds and Tensions," in Shuxun Chen and Charles Wolf, Jr., eds., *China, the United States, and the Global Economy,* Santa Monica, Calif.: RAND, MR-1300-RC, 2001, Chapter Twelve.

Watman, Kenneth, Dean Wilkening, and John Arquilla, *U.S. Regional Deterrence Strategies,* Santa Monica, Calif.: RAND, MR-490-A/AF, 1995.

Weinberger, Caspar, *Fighting for Peace: Seven Critical Years in the Pentagon,* New York: Warner Books, 1991.

White, Thomas E., Secretary of the Army, and General Eric K. Shinseki, *Joint Statement Before the Committee on Appropriations Subcommittee on Defense on the Fiscal Year 2002 Army Budget,* Washington, D.C.: U.S. Senate, June 13, 2001.

Wiseman, Paul, "Big Target Pinned on Small Rebel Group," *USA Today,* February 21, 2002, p. 4.

Woodward, Bob, *The Commanders,* New York: Simon & Schuster, 1991.